T0210674

Contesting Hidden Waters

The world increasingly relies on groundwater resources for drinking water and the provision of food for a growing population. The utilization of aquifer systems also extends beyond freshwater supply to include other resources such as heat extraction and the storage and disposal of substances.

Unlike other books about conflict resolution and negotiations over water resources, this volume is unique in focusing exclusively on conflicts over groundwater and aquifers. The author explores the specific challenges presented by these 'hidden' resources, which are shown to be very different from those posed by surface water resources. Whereas surface watersheds are static, groundwater boundaries are value laden and constantly changing during development.

The book describes the various issues surrounding the governance and management of these resources and the various parties involved in conflicts and negotiations over them. Through firsthand accounts from a pracademic skilled in both process and substance as a groundwater professional and professional mediator, the book offers options for addressing the challenges and issues through a transdisciplinary approach.

W. Todd Jarvis, PhD is Interim Director of the Institute for Water and Watersheds at Oregon State University in Corvallis, Oregon, USA. He is a consulting groundwater hydrologist with nearly 30 years of experience and specializes in dispute prevention and conflict resolution related to groundwater resources and water well construction.

Earthscan Water Text Series

Contesting Hidden Waters
Conflict resolution for groundwater and aquifers
By W. Todd Jarvis

Water Security
Principles, perspectives and practices
Edited by Bruce Lankford, Karen Bakker, Mark Zeitoun and Declan Conway

Water Ethics
A values approach to solving the water crisis
By David Groenfeldt

The Right to Water
Politics, governance, and social struggles
Edited by Farhana Sultana and Alex Loftus

"Anyone involved in water security, development, or even peace building, will treasure this book. Jarvis makes such impressive sense that readers across sectors, disciplines or borders will be inspired to act hand in hand. This work nurtures the idea that various interests can co-exist harmoniously within the context of complex but reasonable transboundary groundwater management."

– Léna Salamé, Programme Specialist, Potential Conflict to Co-operation Potential (PCCP) Coordinator, UNESCO, Paris, France.

"A refreshingly unique and long awaited approach to understanding the most neglected part of the hydrologic cycle and how we interact with it."

– Mark Giordano, Professor and Director of Science, Technology and International Affairs, Georgetown University, USA.

"Population growth brings greater dependency on groundwater. Conflicts are inevitable, and resolution through litigation is often not constructive. Rather than attacking the scientists or the science, Jarvis' Hydro-Trifecta Framework offers a collaborative approach, embracing the best science to facilitate optimum resolutions."

– Steve E. Clyde, Clyde Snow Attorneys, USA.

Contesting Hidden Waters

Conflict resolution for groundwater
and aquifers

W. Todd Jarvis

 Routledge
Taylor & Francis Group

LONDON AND NEW YORK

 earthscan
from Routledge

First published 2014 by Routledge
4 Park Square, Milton Park, Abingdon, Oxon OX14 4RN

605 Third Avenue, New York, NY 10017

Routledge is an imprint of the Taylor & Francis Group, an informa business

British Library Cataloguing in Publication Data
A catalogue record for this book is available from the British Library

Library of Congress Cataloging-in-Publication Data
Jarvis, W. Todd.
Contesting hidden waters : conflict resolution for groundwater and
    aquifers / W. Todd Jarvis.
        pages cm — (Earthscan water text)
    Includes bibliographical references and index.
    1. Groundwater—Management.   2. Aquifers.   3. Conflict
management.   I. Title.
    TD403.J37 2014
    333.91′04—dc23
    2013047751

ISBN: 978-0-415-63292-8 (hbk)
ISBN: 978-0-415-63293-5 (pbk)
ISBN: 978-0-203-09529-4 (ebk)

Typeset in Goudy
by Apex CoVantage, LLC

Contents

Figures and tables

Figures

Tables

Acronyms and abbreviations

APO	Aquifer Protection Overlay
ASR	aquifer storage and recovery
ASTR	aquifer storage, transfer and recovery
CAPP	Casper Aquifer Protection Plan
cfs	cubic feet per second
CGAs	Critical Groundwater Areas
DEQ	Department of Environmental Quality
DLCD	Oregon Department of Land Conservation and Development
DOGAMI	Oregon Department of Geology and Mineral Industries
DSS	decision support systems
EAC	Environmental Advisory Committee
ECR	environmental conflict resolution
EPA	Environmental Protection Agency
EU	European Union
FAO	Food and Agriculture Organization of the United Nations
FoEME	Friends of the Earth, Middle East
GEF	Global Environment Facility
GIS	Geographic Information Systems
gpd	gallons per day
gpm	gallons per minute
GPS	global positioning system
GW	groundwater
IMS	Institute of Portland Metropolitan Studies at Portland State University
IW:Science	International Waters (IW) projects of the Global Environment Facility (GEF)

IWRM	integrated water resources management
IWW	Institute for Water and Watersheds
JFF	joint fact finding
km³	cubic kilometers
KOWSC	Keep Our Water Safe Committee
LUBA	Land Use Board of Appeals
m³pd	cubic meters per day
MGA	Mutual Gains Approach
NGO	nongovernmental organization
NIMBY	not in my backyard
ODR	online dispute resolution
OWRD	Oregon Water Resources Department
ppm	parts per million
RIHN	Research Institute for Humanity and Nature
SADC	Southern Africa Development Community
SDWA	Safe Drinking Water Act
SGO	Sensitive Groundwater Overlay
SIWI	Stockholm International Water Institute
SNWA	Southern Nevada Water Authority
SWIM	Sustainable Water Integrated Management
SWPA	Source Water Protection Area
TaskForce	Umatilla County Critical Groundwater Task Force
TB	transboundary
TOT	time of travel
TCP	tribal consultation policy
UAE	United Arab Emirates
UNESCO	United Nations Educational, Scientific and Cultural Organization
UNESCO-IHE	UNESCO Institute for Water Education
USGS	U.S. Geological Survey
WBPG	Wyoming Board of Professional Geologists
WDEQ	Wyoming Department of Environmental Quality
WDOT	Wyoming Department of Transportation
WHPA	wellhead protection area
WHPP	wellhead protection plan

Acknowledgments

This book started as an accident when I found myself becoming increasingly recruited by decision makers and attorneys on how to protect groundwater from contamination, how to protect aquifers from damage associated with intensive exploitation, and how to share water in areas where water rights exceed actual capacity of groundwater that could be recovered from aquifers. I was educated and trained in the tradition of physical and chemical hydrogeology by one of the most creative thinkers in the field of hydrogeology, Dr. Peter Huntoon. I was like most in the field who knew little about groundwater policy and conflict management. That was messy work for others to do – until things got messy for me.

Leonard and Michelle Hawes at the University of Utah Conflict Resolution Graduate Certificate Program introduced me to the world of conflict resolution. Their patience and tutelage taught me not only how to help others resolve conflict, but also that I was a source of conflict in my work in groundwater hydrology. They provided me with a breath of fresh air, and a new career.

I worked with many Utah-based attorneys over the years as an expert witness, and I am grateful to them for expanding my horizons on how to resolve conflicts over groundwater and innovative ideas on governing groundwater. These include most notably the late Gerald Kinghorn, Steve Clyde, and George Hunt. The state of Utah truly is fortunate to have the wise counsel of these particular water professionals, among many others.

Special thanks go to the colleagues and friends in the state of Wyoming. Many of the individuals I worked with on the wellhead and aquifer protection programs portrayed in this book over the past 20 years were classmates at the University of Wyoming, professional colleagues, dedicated city and county employees, volunteers on advisory committees, as well as courageous city council and county commissioners. I extend my gratitude for permitting me to work shoulder to shoulder with you during my early career as a professional hydrogeologist and fellow citizen-scientist, and during my second career in researching water conflicts. Thank you for welcoming me back to your community and for the time spent on my semistructured interviews and for reviewing the materials that are part of this book.

Internationally renowned mediator Dr. Aaron Wolf introduced me to the field of water conflict transformation and permitted me to work and study under him for over 10 years. Collaborative learning scholar and negotiations expert Dr. Gregg Walker taught me the power of systems thinking and also permitted me to study the nuances of negotiations and collaboration with him over the past 10 years. Both are valued mentors. Online mediation experience with SquareTrade provided valuable insights on the importance of online competency in negotiations. I thank them for training me on the nuances of what I hope leads to increased use of online dispute resolution (ODR) in water conflicts and negotiations.

Dr. Lawrence Susskind of MIT reviewed an early draft of the book proposal and provided many helpful comments and useful guidance on developing the book. Dr. Mark Zeitoun of the University of East Anglia reviewed draft chapters and offered excellent, thought-provoking comments on the value and usefulness of a transdisciplinary approach to water negotiations.

The book cover art is from the River Rock Series of kilnformed and sandblasted glass by Oregon lawyer-turned-artist Sara Morrissey. If one has wondered what groundwater looks like at depths of 500m or more, this artwork captures my observations.

Caryn Davis of Cascade Editing in Philomath, Oregon, provided book author editorial and formatting services. Tim Hardwick and Ashley Wright at Earthscan, Routledge/Taylor & Francis Group, expertly guided me through the process of developing this manuscript. I offer my gratitude to all of these editorial experts for their patience and perseverance in enduring a part-time book writer.

Noted historian, author of many books and filmmaker extraordinaire, chief cajoler to try writing a book of my own, patient manuscript reviewer, continual source of inspiration to me, and wonderful life partner Dr. Kimberly Jensen kept at me to write this book using her patented quota system. It worked. Thanks, Kim.

Preface

Many books and manuals exist on conflict resolution and negotiations for water resources but most focus almost exclusively on surface water with occasional mention of groundwater. No books focus exclusively on conflicts over groundwater resources despite the fact that most of the world's freshwater supplies are underground, that nearly 450 transboundary aquifers have been mapped, and that over 50% of the world's population relies on groundwater for drinking water.

On 9 December 2011 the United Nations General Assembly (UNGA), at its 66th session, adopted the resolution on the 'Law of Transboundary Aquifers' bringing groundwater and aquifers to the fore. This convention is important for many reasons because it recognizes that the utilization of aquifer systems extends beyond just groundwater to include extraction of water, heat and minerals, and storage and disposal of any substance. The conclusion is simple – conflicts over 'hidden waters' are not just about water anymore. Aquifers viewed holistically may have more 'value' than just the water captured from storage.

Most texts on conflict resolution and water negotiations are written by lawyers, academicians, pracademics, and practitioners. This book is different on two accounts: (1) it is written by a hydrogeologist who after 20 years of getting into trouble due to a lack of skills in policy and conflict management, returned to the academies to become enlightened in conflict resolution, negotiations, and dispute prevention to eventually become a professional mediator and instructor; and (2) it is written through the lens of transdisciplinarity. My interests in policy and conflict are focused almost exclusively in groundwater resources, the multiple uses of aquifers, water well construction, and well abandonment. Many of the case studies portrayed in this book are firsthand experiences from my work as a hydrogeologist and mediator.

If the book reads like it is all over the place, welcome to the world of transdisciplinarity. A few years back, I received a peer review on a paper that I prepared with a transdisciplinary focus on groundwater governance and the connection to rivers and submitted to a journal in political geography. The reviewer was openly hostile and without hesitation rejected the paper on the grounds that it read like it was prepared by a schizophrenic. What inspired me to attempt writing about that paper, as well as this book, using the transdisciplinary lens were two important works. The first is a paper by Manfred A. Max-Neef (2005, p5) widely referenced in this book, the first few lines of which state, 'If we go through a list of some of the main problematiques that are defining the new Century, such as water, forced migrations, poverty, environmental crises, violence, terrorism, neo-imperialism, destruction of social fabric, we must conclude that none of them can be adequately tackled from the sphere of specific individual disciplines. They clearly represent transdisciplinary challenges. This should not represent a problem as long as the formation received by those who go through institutions of higher education, were coherent with the challenge. This is, unfortunately, not the case, since uni-disciplinary education is still widely predominant in all Universities'.

The second work that influenced me was Valerie Brown's book *Leonardo's vision: A guide to collective thinking and action*, another wonderful introduction to the notion of the transdisciplinary imagination. As you read my book, you will find that other water professionals have heeded this same calling, but they have also encountered 'rough water' as early career professionals (see Patterson et al., 2013, later in this book). As an avid whitewater river boater, all I can say to these brave young souls is keep the boat straight and both oars in the water as fresh currents are a comin'!

The goal of this book is to encourage water scientists and engineers to integrate conflict management, conflict resolution, dispute prevention and negotiations skills into their transdisciplinary work as it is becoming increasing clear that a mediator's and a facilitator's technical background is critical and directly related to the substance of the dispute. This is especially true for disputes related to groundwater.

References

Brown, V. A., Deane, P. M., Harris, J. A. and Russell, J. Y. (2010) 'Towards a just and sustainable future', in V. A. Brown, J. A. Harris and J. Y. Russell (eds) *Tackling Wicked Problems through the Transdisciplinary Imagination*, Earthscan, Washington, DC, pp1–15

Max-Neef, M. A. (2005) 'Foundations of transdisciplinarity', *Ecological Economics*, vol 53, pp5–16

Patterson, J. J., Lukasiewicz, A., Wallis, P. J., Rubenstein, N., Coffey, B., Gachenga, E. and Lynch, A.J.J. (2013) 'Tapping fresh currents: Fostering early-career researchers in transdisciplinary water governance research', *Water Alternatives*, vol 6, no 2, pp293–312

1 Introduction

A guided journey to the underworld

> On the whole, conflicts over water resources can be linked to environmental, hydro-hegemonic, psychological and ideological, and political and hydro-diplomatic dimensions.
>
> – Ahmed Abukhater (2013)

Conflicts over water are increasing with population growth and changes in climate. Sometimes too much water might not be enough for the variable uses of water to grow food, to grow economies and industry and to grow the environment needed to enrich our spirituality because we do not efficiently store water in soil or underground. Sometimes not enough water might be too much to sustain the machinery of warfare because water is embedded in just about all products from computer chips to fuel.

Conflicts over groundwater and aquifers can be best described as 'wicked' planning problems that have uncertain boundaries, defy absolute solutions and can be a symptom of larger problems (Rittel and Webber, 1973; Innes and Booher, 2010; Ritchey, 2013). Groundwater, or water beneath the Earth's surface, is recognized as a common-pool resource. The management and governance of groundwater resources and the aquifers – underground layers of permeable rock in which accessible quantities of groundwater are found – is challenging and increasingly conflictive.

One cannot directly see groundwater, save for a spring or flowing well. In both developing and developed countries, conflicts can arise due to the plethora of beliefs surrounding the occurrence of groundwater that may be held by various parties – springs and flowing wells are considered by some to be mystical or a divine gift. Groundwater conflicts are more focused on water quality and land use, as opposed to water quantity and allocation conflicts normally associated with surface water disputes. Groundwater boundaries constantly change during development. Conflicts over groundwater include 'identity' issues typically not encountered in surface water conflicts. And conflicts over aquifers can be different from conflicts over the groundwater stored in the aquifers. So groundwater and aquifer situations are 'complex' and typically involve several layers of complexity: multiple parties, multiple issues, deeply held values, important interests, expert knowledge, local knowledge, media and socially embedded conflicts, among many others (Daniels and Walker, 2001; Islam and Susskind, 2013). With the many different types of complexities associated with conflicts over groundwater, there are many competencies that are needed to deal with the conflicts.

The murky and unseen landscape of groundwater and aquifers is not easily navigated without expert guidance. This unseen landscape calls to mind the realm of Charon, the ferryman in Greek mythology, who carried souls of the newly deceased in his crazy boat across the river Styx, which divided the surficial landscape of the living from the underworld landscape, or stygoscape, of the dead. Substance matters, and it now has parity with process and relationships when dealing with conflicts over groundwater and aquifers; this is contrary to a commonly held belief in the conflict resolution field, which is that a mediator's skills are more important than the mediator's knowledge of the subject matter. However, specialization in mediation and conflict resolution is increasingly recognized as essential. In a 2010 video lecture on Mediate.com, international negotiations expert Lawrence Susskind (2010) describes why specializations within mediation are necessary. He proposes that the mediator should understand not only the process, but also the substance in order to offer options and understand the language within the institutional context. Kane (2012), who provides an interesting case study on a conflict over fugitive stormwater, offers the following: 'When coming to agreement on the technical solution was the predominant interest, the mediator's technical background was critical and related directly to the substance of the dispute'. In 2012, the Michigan State Legislature passed Public Act 602, Aquifer Protection and Dispute Resolution, which acknowledges the challenges for resolving disputes between small-quantity groundwater users, such as domestic uses, versus larger volume users in the agriculture industry. In Australia, Dempsey (2013) reports on a poll of in-house dispute resolution counsel from 76 very large international corporations, where 85% relied upon the mediator as having expertise in the core issue of the case when determining whom to select. I foresee enormous opportunities in niche groundwater diplomacy, building off of the success of the Singapore's experience in niche water diplomacy (see Jarvis, 2013).

Renevier and Henderson (2002) indicate that conflicts over water require a holistic approach to address multidisciplinary and multimedia issues. I would go one step further and say groundwater conflicts require tackling this wicked problem through the transdisciplinary imagination, as described by Brown and others (2010, p5). They define the transdisciplinary approach 'to be the collective understanding of an issue; it is created by including the person, the local and the strategic, as well as specialized contributions of knowledge. This is distinguished from multidisciplinary inquiry, which is taken to be a combination of specializations to a particular purpose, and from interdisciplinary, the common ground between two specializations that may develop into a discipline of its own. Transdisciplinary approaches require the use of imagination typically associated with creativity, insight, vision and originality and it is also related to memory, perception and invention'. In his wonderful collection of short stories about groundwater, hydrologist Francis Chapelle (2000, p50) underscores the importance of imagination for groundwater conflicts by writing 'without imagination, both hydrologists *and* water witches would be blinded. After all, mysticism is as much a child of human imagination as is rationality'. And Patterson and others (2013) consider the transdisciplinary approach as critical to research for the rapidly evolving world of water governance, even proposing that the approach may lead to 'inventing new science'.

What are the main causes that usually lead to water conflicts?

The answer to this question is 'it depends'. Delli Priscoli and Wolf (2009) invite the practitioner to use a five-piece Circle of Conflict pie chart to diagnose causes of conflict.

1. Relationships (poor communication, negative behavior)
2. Data (interpretation, misinformation, procedures)
3. Interests (perceived competition, procedural interests)
4. Structural (unequal power in terms of bargaining, material and ideational power, time, destructive behavior, geography)
5. Values (ideology, spirituality)

In my experience as a hydrogeologist and mediator, the issue of identity as a cause for conflict, described by Rothman (1997), is especially unique to groundwater and aquifers, as discussed in the case studies in subsequent chapters. So the groundwater conflict pie has a sixth piece – Identity (history, control, dueling experts, folk beliefs).

But there is a plethora of other reasons for conflicts over groundwater and aquifers that has as much to do with history as with the hidden nature of the resources. The following sections are overviews of the causes of conflicts that I consider unique to groundwater and aquifers. The reader may become as confused and exasperated by the listing as policy and decision makers who must work with groundwater professionals.

Multiple working hypotheses

Conflicting conceptual models are part of the technical training of hydrogeologists focusing on the intellectual method of 'multiple working hypotheses' introduced in the late 1890s by the first hydrogeologist in the United States, Thomas Chamberlain (Chamberlain, 1897). The structure of the method of multiple working hypotheses revolves around the development of several hypotheses to explain the phenomena under study. The antithesis of multiple ways of knowing is considered a 'ruling theory', as described by Wade (2004).

One of the best documented examples of multiple working hypotheses in play focuses on the Santa Cruz Aquifer shared between the United States and Mexico. Milman and Ray (2011) carefully document the problems of inadequate information (what they refer to as 'epistemic uncertainty') that led to framing uncertainty or ambiguities in how the situation was understood and represented. The U.S.-Mexico water agencies apparently operate independently; coordinated data collection and sharing does not exist. Both countries hold different conceptual models of the aquifer, leading to different interpretations of water availability, impacts of groundwater use, recharge and protection activities. Both countries agree to a conceptual model of the aquifer composed of three hydrogeologic layers, yet the U.S. model is of shallower, less conductive layers, whereas the Mexico model depicts a deeper aquifer with great conductivity. The model developed for the Mexican side of the border estimates recharge rates nearly five times greater than the U.S. model. The U.S. water values center on protecting nature and the character of the border community. Mexico's water management values meeting human needs and prioritizing future growth and development. The different definitions of sustainable yield and the culture of water held by each country are reflected in their groundwater development plans and practices.

Milman and Ray (2011) provide good advice regarding conflicting conceptual models by suggesting that negotiations might best begin through efforts designed to reduce multiple uncertainties that exist, rather than seeking cooperative solutions that maximize joint benefits or commitments to allocations of groundwater. They wisely suggest moving toward 'low-regret' steps that allow for adaptive management objectives.

Junk science

The extreme antithesis of the multiple working hypotheses and the scientific method has many references, but perhaps the most used and noteworthy term is 'junk science'. There are many definitions, but my favorite is by two groundwater hydrologists, Allan Freeze and Jay Lehr (2009, p300). It is not unique to groundwater, but is often prevalent in debates about groundwater and aquifers because the distribution of data and information is often less than ideal, due to time and expense. Freeze and Lehr describe junk science the following way:

> 'the mirror image of real science, with much of the same form, but none of the same substance' (Huber, 1991). Junk science features biased data, spurious inference, wishful thinking and logical legerdemain. It is often carried out in support of political goals. The inferences that are drawn from the scientific data are usually self-serving. Unexplained correlations of related parameters may be presented as if they were proven cause-and-effect relationships. Or it may be argued that whatever has not been proven false must be true. The junk scientist is good at 'counting the hits and forgetting the misses' (Sagan, 1996). Those who put forward shady science in support of political goals have consciously or unconsciously given priority to their doctrinaire beliefs over their scientific responsibilities.
>
> (Freeze and Lehr, 2009, p300)

Conversely, Pielke (2007, p126) defines 'sound science' as meaning 'that one believes that political agendas following from that science are right, just, and deserving of support'. Pielke indicates that battles often take place over 'whether science is sound or junk instead of debating the value or practicality of specific policy alternatives'. We will learn more about this debate in case studies presented later in this book.

Stealth issue advocacy

Hydrogeologic information is key in many public policy debates regarding ground-water use and impacts. As geoscientists, we have earned a seat at the table in environmental policy debates. And while geoscientists strive to complete hydro-geologic work in an objective manner, using the best available science, some of the 'science' is being presented with shaky foundations, namely, to bolster personal policy preferences. Prominent Pacific Northwest fisheries biologist and colleague Robert Lackey (2007) refers to this as normative science, and defines it as 'information that is developed, presented or interpreted based on an assumed, usually unstated, preference for a particular policy choice'. Lackey calls it stealth advocacy 'because the average person reading or listening to such scientific state-ments is likely unaware that it is a form of advocacy. Normative science is a corruption of science and should not be tolerated in the scientific community – without exception' (Lackey, 2007, p17).

Pielke (2007, p62) states that scientists have choices about how they engage the broader society of which they are part, and that hiding behind science is not a productive option because it risks damaging the positive contributions of their own expertise. What is the potential result? 'Because science is highly valued as a source of reliable information, disputants look to science to help legitimate their interests. In such cases, the scientific experts on each side of the controversy effectively cancel each other out, and the more powerful political or economic interests prevail, just as they would have without the science'.

During the course of listening to a former congressman give a lecture at a major U.S. university, I asked how scientists could make more of an impact in policy making and politics. His response? Scientists should become politicians and par-ticipate in the bargaining, negotiation and compromise associated with politics. Some geoscientists have made the jump from science to politics, most notably former Secretary of the Interior Bruce Babbitt (geophysicist), U.S. Senator Har-rison Schmitt (astronaut and geologist) and Oregon senator Jackie Dingfelder (land use planner).

But for rank and file groundwater professionals, the message is this: it is important to get involved in policy deliberations, but remember that our job as geoscientists in getting the message across to lay people is already stacked against us, as we work in mostly a 'hidden' environment that is not seen by the public and is often con-sidered mystical. The line between hydrogeology and stealth-issue advocacy can be a fuzzy one, but the damage to the profession is crystal clear if the line is crossed haphazardly.

Tensions between the political and the technical

According to Delli Priscoli (2004), few issues mesh the technical and political as much as water management and governance. The traditional model assumes a sep-aration between the political, typically associated with legislative action, and the technical, typically associated with implementing executive agencies. Delli Priscoli (2004) suggests that complex water-management decisions often break down the distinction between the two factions, and that the implementation and administration of laws make the distribution of impacts clearer. It is during implementation that the political benefits become clear, and when administrators of technical agencies 'begin to appear as bestowers or deniers of political benefits' (Delli Priscoli, 2004, p224).

I experienced this tension while assisting a community located in the northwestern United States that was proactive about addressing a serious groundwater depletion problem – water level declines approaching 150 m over the past 50 years – given that the area was nearly 100% dependent on groundwater for irrigation and municipal drinking water supplies. Even before the draft water management plan was completed, the first source of tension among some of the 'science-based' stakeholders focused on the conceptual hydrogeologic model of the deep basalt aquifers. In concert with local hydrologists, I synthesized the water level data maintained by the state water resources program and integrated these data with the geologic data compiled by the state geologic survey. The state water resources department disagreed with the conceptual model, indicating that it was too simplistic, yet hesitated to offer an alternative conceptual model for reasons that remained unsaid, but were more than likely linked to politics, since the aquifer is shared with the state of Washington (the neighboring state to the north), and the Confederated Tribes of the Umatilla Indian Reservation. The conceptual model could have been perceived as 'outside the current paradigm', as described by Shomaker (2007), who was proposing a new groundwater development scheme in New Mexico.

Hydromyths and the hydrohydra

There is a prevalence of misinformation or 'hydromyths', even among water experts and resource managers, that has infiltrated policy and fueled conflicts over groundwater and aquifers. R. Llamas and Custodio (2002, p10) indicate two prominent hydromyths: (1) that groundwater is an unreliable or fragile resource where 'almost every well becomes dry or brackish after a few years' and (2) that groundwater mining (or development of nonrenewable groundwater resources) is always unethical because it is unsustainable and damages future generations. There also exists a 'hydrohydra' – a many-headed beast – of myths, paradoxes and misunderstandings of the tenets within hydrogeology that ultimately leads to a lack of trust by decision makers.

Hydromyths permeate the debate over the use and misuse of groundwater resources that can be tied to the misinterpretation of some of the fundamental precepts in groundwater hydrology. One of the most egregious is the concept of safe yield. The safe yield concept is developed around the Law of Conservation of Matter – the reasoning that if an aquifer is pumped at a rate that exceeds the rate of recharge, the aquifer will be depleted (Devlin and Sophocleous, 2005). The recognition of this hydromyth did not receive attention by water experts until 1997, when Sophocleous (1997) wrote an editorial entitled, 'Managing Water Resources Systems: Why "Safe Yield" Is Not Sustainable' for *Groundwater*, the international journal published by the National Ground Water Association.

Bredehoeft (1997, p929) echoed the problem regarding this notion in a follow-up editorial entitled 'Safe Yield and the Water Budget Myth'. Bredehoeft reports that 'sustainable groundwater developments have almost nothing to do with recharge'.

But groundwater planners and experts joust with a hydrohydra of issues associated with developing and protecting groundwater that are distributed *within* the field of hydrology. Greek mythology describes a nine-headed water beast that dwelled in a marsh near Lerna, Greece; Hercules was sent to kill the serpent as the second of his 12 labors. 'Hydra' is a term applied in this chapter to the complex situations or problems that continually pose compounding difficulties in groundwater governance. The safe-yield controversy is just one head of the hydra of issues that fuels the differences in how dueling experts define and prioritize groundwater development concerns. As depicted in Figure 1.1, the dueling experts joust with a hydrohydra of issues associated with governance.

In many parts of the world, groundwater is part of the public trust and is a classic 'common pool resources' – a resource that is available to all, difficult to exclude users from gaining access to and is 'extractable', becoming unavailable to other users after some of the resource has been used or 'mined' (Adams et al., 2003;

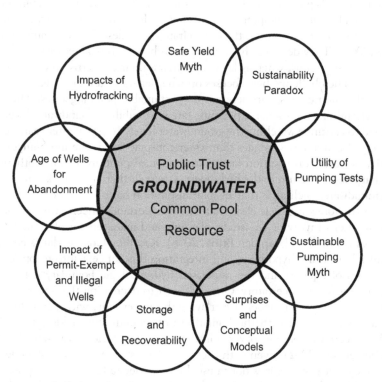

Figure 1.1 Hydrohydra, of myths, paradoxes and misunderstandings of the tenets within hydrogeology

Margat and van der Gun, 2013). The objective of safe-yield calculations is to determine a 'single-product exploitation goal' or the quantity of groundwater that can be pumped year after year without destroying the resource base – in other words, the 'sustainability' of an aquifer (Sophocleous, 1997). Yet the paradox of sustainability as it relates to groundwater resources is that this concept has traditionally focused on sustainability as it relates to the objectives of society, now and in the future (Alley and Leake, 2004; Maimone, 2004). Until recently, little consideration has been paid to the sustainability of ecosystems that also rely on groundwater. Sophocleous (1997, p561) indicates that the determination of a fixed sustainable yield for an aquifer is not possible because 'yield varies over time as environmental conditions vary'. Alley (2001) indicates that little long-term monitoring of groundwater systems responsiveness to climate variability exists. Alley and Leake (2004, p13) suggest that safe yield and sustainability are value-laden concepts and '(are) in the eye of the beholder'.

The intractable arguments about safe yield and sustainability, as related to groundwater resources, are often ignored by practitioners, but there are many other hurdles to overcome due to the hydrohydra. Given that water resources managers are searching for figures for long-term planning in order to then advise farmers, industries and landowners considering large investments in land development, there is pressure on the groundwater experts to provide an opinion on the sustainable rate of pumping from an aquifer. Here, groundwater experts could serve as a solution to conflict through reframing, as described by Kaufman and Gray (2003). They describe the differences in how technical experts frame risk in probabilistic terms, whereas lay people frame their concerns on catastrophic extremes. The challenge here focuses on whether water use or water consumption is the metric to use in determining sustainable pumping rates. The goal with determining a sustainable pumping rate is to stabilize groundwater levels, as opposed to continue watching the groundwater levels drop year after year. Kendy (2003, p3) convincingly argues that water consumption is the more important metric, as water consumption factors in evapotranspiration or water otherwise lost that does not return to the hydrologic system; furthermore, the key to arresting groundwater declines is that 'evapotranspiration has to decrease'.

From the foregoing, it is clear that there is enormous uncertainty associated with situations involving the study, planning and management of groundwater resources (Tidwell and van den Brink, 2008). Recalling Chamberlain's concept of multiple working hypotheses and integrating this concept into mapping the boundaries of aquifers and the associated hydrogeologic boundary conditions yields one level of uncertainty. One approach to decreasing the uncertainty is the use of groundwater models. Overlaying this uncertainty with the hydrologic boundaries associated with the human use of groundwater yields yet another level of uncertainty because of the transient character of the capture-area boundary (Bredehoeft, 2002). Even with the use of sophisticated numerical modeling of groundwater systems, it is not unusual to experience failures or 'surprises' in 25% to 30% of the groundwater models and associated hydrologic boundaries (Bredehoeft, 2005). Ozawa (2005) indicates that any system developed for folding science and other sorts of information into decision making must also be able to make space for silent voices and multiple ways of knowing.

Many times, groundwater experts will avoid the safe-yield, sustainability and sustainable pumping issues altogether by estimating the useful life of an aquifer – by dividing an estimate of recoverable water in storage by an estimate of annual water consumptive use (Alley, 2006). Obvious uncertainties with such an approach include the values used in the numerator and denominator; more importantly, it is impossible to remove all water stored in an aquifer with pumping wells (Alley, 2006). Likewise, the depletion of the aquifer using this approach ignores the potential effects associated with land subsidence and the water needs of ecosystems. Yet water resource planning agencies often hear groundwater experts extolling the virtues of relying on groundwater resources, due to the immense quantity of water stored underground, without reference to what quantity can be realistically recovered.

The spectacular increase in groundwater use has been mainly driven by economic reasons: the full direct cost of groundwater use is usually a small fraction of the value of crops obtained with groundwater abstraction; likewise, groundwater use in the development of rural subdivisions is also a small fraction of the value of the housing dependent on groundwater abstraction. In both agriculture and during the course of developing 'sagebrush subdivisions', large exurban areas of growth are typically remote from water and wastewater services. This situation has become what M. Llamas and Martínez-Santos (2005a, 2005b) refer to as a 'silent revolution'. With the silent revolution come problems such as the degradation of groundwater quality, excessive drawdown of groundwater levels, land subsidence, reduction in spring flow and base flow and groundwater dependent ecosystems. But the silent revolution also leads to the strengthened capacity of groundwater beneficiaries, such as farmers, real estate developers, well drillers and landowners, to form political lobbies with political clout. Yet the apparent disconnect between land use and water use is building surprises and consigning public priorities to resources that may experience death by a thousand cuts (Van de Wetering, 2007).

Water well drillers and other groundwater experts often promote the concept that wells can last for generations. Well drillers are understandably proud of their work, and well design engineers are understandably nervous about errors and omissions that increase the likelihood of liability claims and associated insurance premiums. The debate often focuses on the life expectancy of a well and who pays for end-of-life issues. Ownership of abandoned, corroded and unused wells is a liability issue when cross communication, surface water interference or contamination of municipal water supplies occurs. The landowner, well driller, state regulatory agencies and municipalities may often be placed in situations where the only winners are 'conflict beneficiaries' – the attorneys and their expert witnesses.

Jarvis and Stebbins (2012) ascertained that the design life of a typical well in reality is 20 to 30 years, due to changes in regulatory design standards, corrosion and contamination, as well as fluctuations in water levels due to well interference,

groundwater mining and climate change. Steichen et al. (1988) found that factors correlating with nitrate contamination in water wells included the age of
the well, the land use around the well and the distance to a possible source of
contamination. Old water wells were more likely to be contaminated than new
wells because the age of the well is an indicator of the history of land use near the
well, as well as the length of time that potentially polluting activities have been
located in the vicinity of the well. Their statistical analyses found a strong cor-
relation between nitrate contamination and wells that were about 30 years old.

The rapid expansion of installing directionally drilled boreholes in regionally
extensive shales and low permeability ('tight') sandstones, followed by staged
stimulation through high pressure injection of fluids in order to induce fracturing,
or 'fracking', to recover natural gas, is a media darling that is currently fueling
public angst and concern, primarily about water supplies and groundwater pollution. Documentary films such as *Gasland*, *FrackNation* and *Gasland, Part II*
are new media examples focusing on the debate (see Appendix B). Jurisdictions
across the globe (New York State, Quebec [Canada], Germany, the Netherlands
and France) are so concerned about the uncertainties associated with fracking
and groundwater contamination that moratoria are in place until the media storm
and uncertainties subside and are rectified (Jackson et al., 2013). The debate
among groundwater professionals extends from the concern that fracking causes
earthquakes in Ohio (Schwartz, 2013), to whether the fracking process may
propagate fractures to the ground surface, to the role of fugitive gas leakage and
fracking fluids from active hydrocarbon wells, abandoned wells, natural faults and
fissures (Jackson et al., 2013). On the basis of peer-reviewed studies completed
in disparate geologic regimes, Johnson (2013) posits the question 'Does fracking
contaminate groundwater?' ultimately arriving at the conclusion that there is not
a yes or no answer.

The dilemma of boundaries

Defining boundaries around water resource domains is 'a supremely political
act' because boundaries represent different interpretations of key issues, such
as water quality, water quantity, nature, economics and history (Blomquist and
Schlager, 2005). I was invited to participate in an international symposium titled
'The Dilemma of Boundaries: Toward a New Concept of Catchment' hosted
by the Research Institute for Humanity and Nature (RIHN) in Kyoto, Japan
in 2009, where it was interesting to see that all that falls within the confines of
the boundary has a common bond. The existence of a boundary is the criterion
for the individuality of an autonomous entity. Boundaries define who is in, who
is out; what is permissible, what is not; what needs to be protected and what is
already protected. Yet with the assumptions associated with the rights and implied
boundaries comes the fact that the assumptions, knowledge and understandings
that underlie the definition of the rights and associated boundaries are uncertain
and often contested (Adams et al., 2003). The uncertain and contested boundaries associated with groundwater and aquifers were the topics of my contribution

(Jarvis, 2012). The following is an updated summary of why boundaries matter in the conflicts over groundwater and aquifers.

Few political geographers have addressed the problem of how boundaries are placed around common pool resources such as groundwater. I often read that the boundaries of groundwater and aquifers are 'fuzzy' because of the vagaries in where recharge areas are located, the hydrologic connection to surface water resources and flow and discharge characteristics that are known at only a reconnaissance level (Theesfeld, 2010).

Yet the literature is replete with boundaries for groundwater domains. A consideration of transdisciplinary approach to exploring the geopolitics of ground-water yields a typology for groundwater and aquifer boundaries. As depicted in Figure 1.2, my work found that (1) traditional approaches to defining ground-water domains focus on predevelopment conditions, including the 'domes' and 'veins' defined through folk beliefs, referred to herein as a bona-fide 'commons' boundary; (2) groundwater development creates a human-caused or fiat 'hydro-commons' boundary, where hydrology, hydraulics, property rights and economics

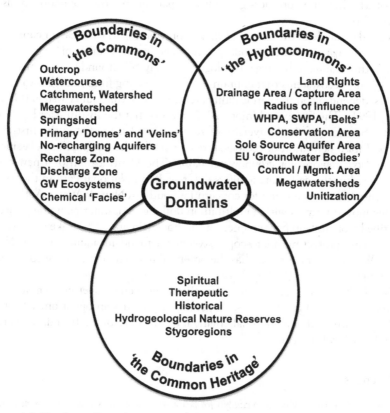

Figure 1.2 Typology of boundaries in groundwater domains

are meshed; and (3) the social and cultural values of groundwater users define a fiat 'commons heritage' boundary, acknowledging that groundwater resources are part of the 'common heritage of humankind'. The significance of this typology is that it focuses more on 'problemsheds' and 'policysheds' – the boundaries of a particular problem or policy defined by the groundwater and aquifer users – rather than just on watersheds or catchments as proffered by integrated water resources management (Foster and Ait-Kadi, 2012).

The new world order of groundwater will focus on the delineation of resource and user-domain boundaries, regardless of whether (1) the technological options to manage groundwater quantity and quality problems employ water transfers, managed recharge or conjunctive use; or (2) the resource governance solutions include (a) collective or community action, (b) developing instrumental approaches such treaties, agreements, rights, rules and (c) prices or other incentives, such as preserving the structural and ecological integrity of groundwater systems, as summarized by Giordano (2009). As a consequence, politics dictates a fairly well-defined boundary that is connected to a desired outcome. These desired outcomes need to be spelled out in the new instruments for groundwater and aquifers, as the boundaries are an important part of governance, regardless of whether one considers groundwater part of the 'the commons' or as 'a commodity'.

A good example of the changing values contributing to what can or cannot be done within a groundwater boundary is the emerging importance of 'stygoregions' within the common heritage typology. Virtually all groundwaters constitute underground ecosystems that contain life forms ranging from microbes to larger species, depending on the permeability architecture of the aquifer system (Gibert et al., 1994). The European Union (2012) reports that the Institute for Environmental Sciences of the University of Koblenz-Landau presented the first-ever proposal for a definition of ecological references for the geographical classification of 'stygoregions' for groundwater fauna. The identification of a new cave species within the carbonate aquifers of Great Basin National Park in the western United States was cause for concern by Southern Nevada Water Authority, as they moved forward with a multibillion dollar groundwater pumping project, described more fully in later chapters of this book (Brean, 2009). Likewise, 'water trigger' laws protecting microscopic stygofauna found in aquifers in the New South Wales, Australia region has the potential to derail a large-scale coal-seam gas-extraction project (Cubby, 2013).

Clearly, boundaries can create competition between competing communities of groundwater and aquifer users. Appendix A provides an opportunity for the reader to learn firsthand how challenging it can be to resolve a boundary-related issue through role playing.

References

Abukhater, A. (2013) *Water as a catalyst for peace – transboundary water management and conflict resolution*, Routledge, Oxon, UK

Adams, W.M., Brockington, D., Dyson, J., and Vira, B. (2003) 'Managing tragedies: Understanding conflict over common pool resources', *Science*, vol 302, pp1915–1916

Alley, W.M. (2001) 'Ground water and climate', *Groundwater*, vol 39, no 2, p161

Alley, W.M. (2006) 'Another water budget myth: The significance of recoverable ground water in storage', *Groundwater*, vol 45, no 3, p251

Alley, W.M. and Leake, S.A. (2004) 'The journey from safe yield to sustainability', *Groundwater*, vol 42, pp12–16

Blomquist, W.A. and Schlager, E. (2005) 'Political pitfalls of integrated watershed management', *Society and Natural Resources*, vol 18, pp101–117

Brean, H. (2009) 'New cave species have been identified at Great Basin National Park', *Las Vegas Review-Journal*, 4 October, www.reviewjournal.com/news/new-cave-species-have-been-identified-great-basin-national-park

Bredeheoft, J.D. (1997) 'Safe yield and the water budget myth', *Groundwater*, vol 35, no 6, p929.

Bredeheoft, J.D. (2002) 'The water budget myth revisited: Why hydrogeologists model', *Groundwater*, vol 40, no 4, pp340–345

Bredehoeft, J.D. (2005) 'The conceptualization model problem – surprise', *Hydrogeology Journal*, vol 13, no 1, pp37–46

Brown, V.A., Deane, P.M., Harris, J.A. and Russell, J.Y. (2010) 'Towards a just and sustainable future', in V.A. Brown, J.A. Harris and J.Y. Russell (eds) *Tackling wicked problems through the transdisciplinary imagination*, Earthscan, Washington, DC, pp1–15

Chamberlin, T.C. (1897) 'The method of multiple working hypotheses', *Journal of Geology*, vol 5, pp837–848

Chapelle, F.H. (2000) *The hidden sea: Groundwater, springs, and wells*, National Ground Water Association, Westerville, OH

Cubby, B. (2013) 'The bug that bit Santos' $1 billion gas project', *The Sydney Morning Herald*, 12 July, www.smh.com.au/environment/conservation/the-bug-that-bit-santos-1-billion-gas-project-20130711-2pt6x.html#ixzz2aHTEds2y

Daniels, S.E. and Walker, G.B. (2001) *Working through environmental conflict: The collaborative learning approach*, Praeger, Westport, CT

Delli Priscoli, J. (2004) 'What is public participation in water resources management and why is it important', *Water International*, vol 29, no 2, pp221–227

Delli Priscoli, J. and Wolf, A. (2009) *Managing and transforming water conflicts*, Cambridge Press, New York, NY

Dempsey, T. (2013) 'Australia: The times are a changing – when it comes to selecting a mediator', *Mondaq*, Australia, Litigation, Mediation & Arbitration, www.mondaq.com/australia/x/237632/Arbitration+Dispute+Resolution/The+times+are+a+changing+when+it+comes+to+selecting+a+mediator

Devlin, J.F. and Sophocleous, M. (2005) 'The persistence of the water budget myth and its relationship to sustainability', *Hydrogeology Journal*, vol 13, pp549–554

European Union (2012) 'Groundwater protection should be established in law', 10 October, *Balkans.com Business News*, www.balkans.com/open-news.php?uniquenumber=161025

Foster, S. and Ait-Kadi, M. (2012) 'Integrated Water Resources Management (IWRM): How does groundwater fit in?' *Hydrogeology Journal*, vol 20, pp415–418

Freeze, R.A. and Lehr, J.H. (2009) *The fluoride wars: How a modest public health measure became America's longest-running political melodrama*, John Wiley & Sons, Hoboken, NJ

Gibert, J., Stanford, J.A., Dole-Oliver, M.J. and Ward, J.V. (1994) 'Basic attribute of groundwater ecosystems and prospects for research', in J. Gibert, D. Danielpol and J. Stanford (eds) *Groundwater Ecology*, Academic Press, Inc., San Diego, CA, pp7–40

Giordano, M. (2009) 'Global groundwater? Issues and solutions', *Annual Review of Environment and Resources*, vol 34, pp7.1–7.26

Innes, J. E. and Booher, D. E. (2010) *Planning with complexity: An introduction to collaborative rationality for public policy*, Routledge, New York, NY

Islam, S. and Susskind, L. E. (2013) *Water diplomacy: A negotiated approach to managing complex water networks*, RFF Press, Routledge, New York, NY

Jackson, R. E., Gorody, A. W., Mayer, B., Roy, J. W., Ryan, M. C. and Van Stempvoort, D. R. (2013) 'Groundwater protection and unconventional gas extraction: The critical need for field-based hydrogeological research', *Groundwater*, vol 51, no 4, pp488–510

Jarvis, W. T. (2012) 'Integrating groundwater boundary matters into catchment management', in M. Taniguchi and T. Shiraiwa (eds) *The Dilemma of Boundaries: Toward a New Concept of Catchment*, Global Environmental Studies, Springer, Tokyo, pp161–176

Jarvis, W. T. (2013) 'Water scarcity: Moving beyond indices to innovative institutions', *Groundwater*, vol 51, no 5, pp663–669

Jarvis, W. T. and Stebbins, A. (2012) 'Examining exempt wells: Care for exempt wells provides opportunities for the water well industry', *Water Well Journal*, September, pp23–27

Johnson, S. K. (2013) 'More evidence for (and against) groundwater contamination by shale gas', *Ars technica*, 28 June, http://arstechnica.com/science/2013/06/more-evidence-for-and-against-groundwater-contamination-by-shale-gas/

Kane, C. (2012) 'Mediating storm water run-off disputes – settling "neighbor wars"', *Mediate.com*, www.mediate.com/articles/KaneC1.cfm, last accessed 16 December 2012

Kaufman, S. and Gray, B. (2003) 'Retrospective and Prospective Frame Elicitation', in R. O'Leary and L. B. Bingham (eds) *The Promise and Performance of Environmental Conflict Resolution*, Resources for the Future, Washington, DC, pp129–147

Kendy, E. (2003) 'The false promise of sustainable pumping rates', *Groundwater*, vol 41, no 1, pp1–4

Lackey, R. T. (2007) 'Science, scientists, and policy advocacy', *Conservation Biology*, vol 21, no 1, pp12–17

Llamas, M. R. and Martínez-Santos, P. (2005a) 'Intensive groundwater use: Silent revolution and potential source of social conflicts', *American Society of Civil Engineers Journal of Water Resources Planning and Management*, vol 131, no 4, pp337–341

Llamas, M. R. and Martínez-Santos, P. (2005b) 'The silent revolution of intensive ground water use: pros and cons', *Groundwater*, vol 43, no 2, p161

Llamas, R. and Custodio, E. (2002) 'Intensively exploited aquifers – main concepts, relevant facts, and some suggestions', Series on Groundwater No. 4, UNESCO, IHP-VI

Maimone, M. (2004) 'Defining and managing safe yield', *Groundwater*, vol 42, no 6, pp809–814

Margat, J. and van der Gun, J. (2013) *Groundwater around the world: A geographic synopsis*, CRC Press/Balkema, Leiden, The Netherlands

Milman, A. and Ray, I. (2011) 'Interpreting the unknown: Uncertainty and the management of transboundary groundwater', *Water International*, vol 36, no 5, pp631–645

Ozawa, C. (2005) 'Putting science in its place', in J. T. Scholz and B. Stiftel (eds) *Adaptive Governance and Water Conflict: New Institutions for Collaborative Planning*, Resources for the Future, Washington, DC, pp185–195

Patterson, J. J., Lukasiewicz, A., Wallis, P. J., Rubenstein, N., Coffey, B., Gachenga, E. and Lynch, A. J. J. (2013) 'Tapping fresh currents: Fostering early-career researchers in transdisciplinary water governance research', *Water Alternatives*, vol 6, no 2, pp293–312

Pielke, R. A. (2007) *The honest broker: Making sense of science in policy and politics*, Cambridge University Press, Cambridge, UK

Renevier, L. and Henderson, M. (2002) 'Science and scientists in international environmental negotiations', in L. Susskind, W. Moomaw and K. Gallagher (eds) *Transboundary Environmental Negotiation – New Approaches to Global Cooperation*, Jossey-Bass, A Wiley Company, San Francisco, CA, pp107–129

Ritchey, T. (2013) 'Wicked problems: Modelling social messes with morphological analysis', *Acta Morphologica Generalis*, vol 2, no 1, pp1–8

Rittel, H. and Webber, M.M. (1973) 'Dilemmas in a general theory of planning', *Policy Sciences*, vol 4, pp155–169

Rothman, J. (1997) *Resolving identity-based conflict in nations, organizations and communities*. Jossey-Bass, A Wiley Company, San Francisco, CA

Schwartz, F.W. (2013) 'Folk beliefs and fracking', *Groundwater*, vol 51, no 4, p479

Shomaker, J. (2007) 'What shall we do with all of this groundwater?' *Natural Resources Journal*, vol 47, no 4, pp781–791

Sophocleous, M. (1997) 'Managing water resources systems: Why "safe yield" is not sustainable', *Groundwater*, vol 35, no 4, p561

Steichen, J., Koelliker, J., Grosh, D., Heiman, A., Yearout, R. and Robbins, V. (1988) 'Contamination of farmstead wells by pesticides, volatile organics, and inorganic chemicals in Kansas', *Ground Water Monitoring & Remediation*, vol 8, no 3, pp153–160

Susskind, L. (2010) 'Larry Susskind: Specializations in mediation as essential – video', www.mediate.com/articles/susskinddvd04.cfm, last accessed 16 December 2012

Theesfeld, I. (2010) 'Institutional challenges for national groundwater governance: Policies and issues', *Groundwater*, vol 48, no 1, pp131–142

Tidwell, V.C. and van den Brink, C. (2008) 'Cooperative modeling: Linking science, communication, and ground water planning', *Groundwater*, vol 46, no 2, pp174–182

Van de Wetering, S.B. (2007) 'Bridging the governance gap: Strategies to integrate water and land use planning', Collaborative Governance Report No. 2, Public Policy Research Institute, the University of Montana

Wade, J.H. (2004) 'Dueling experts in mediation and negotiation: How to respond when eager expensive entrenched expert egos escalate enmity', *Conflict Resolution Quarterly*, vol 21, no 4, pp419–436

2 Groundwater governance versus aquifer governance

The chasm between formal hydrology and popular hydrology [is] how we scientists would like things to be and what people actually want. . . . To create a platform for scientists and farmers to work together, it is critical that these two world views merge. . . . The core groundwater governance challenge, then differs from one groundwater socioecology to another.

– Tushaar Shah, *Taming the Anarchy* (2009)

When it comes to governing or managing groundwater resources, what is it that one is really trying to govern – the groundwater stored and captured by wells and drained by springs, or the 'container', the aquifer that stores groundwater? This chapter introduces the reader to both approaches of governance and introduces a transdisciplinary approach to governing the subterranean resources that links communities of users to approaches of increasing competition and accountability, takes a look forward toward exploring for new aquifers, takes a look forward at future uses of depleted aquifers and uses diplomacy to share lessons learned from the transdisciplinary imagination.

Eight to 10 million cubic kilometers (km^3) of freshwater constitute our 'hidden sea' of groundwater (Margat and van der Gun, 2013). The number is staggering to conceptualize, so, for comparison, consider Lake Baikal in Russia: it is both the deepest lake in the world and the largest by volume of freshwater at 23,600 km^3 – more than the combined volume of all of the Great Lakes of North America. Earth's groundwater resources could fill nearly 425 Lake Baikals.

Groundwater is the world's most extracted raw material, with withdrawal rates estimated to range from 800 to 1,000 km^3 per year through millions of water wells (Shah, 2009; Margat and van der Gun, 2013). Groundwater is everywhere, and the water quality is such that many times it is available at the point of use, as opposed to surface water, which may have to be diverted for long distances and may require treatment before it can be used.

Groundwater is also the 'reservoir of choice', as it is not affected by droughts as much as surface water resources are; the large amount of storage afforded in aquifers increases the resiliency of this water resource. Yet despite its volume and resiliency, few indices of water scarcity listed in Figure 2.1 incorporate

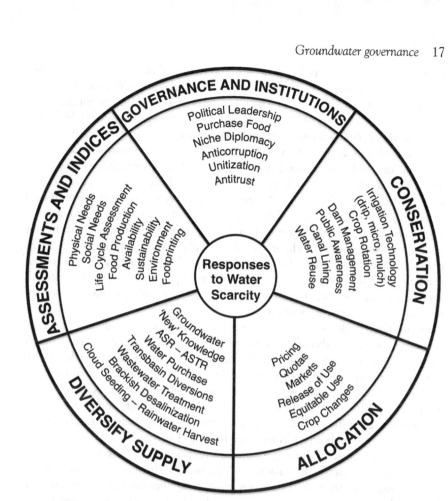

Figure 2.1 Responses to water scarcity; from Jarvis (2013)

groundwater into the assessments. The early emphasis of water scarcity research was largely directed toward surface water, as the quantity of surface water resources are known with a reasonable degree of certainty. Groundwater has become more integral to the analyses, however, nearly 20 years after the introduction of the water scarcity conceptual framework (Jarvis, 2013).

Given the volume of 'fresh' groundwater coupled with the current rates of extraction, groundwater could theoretically provide water supplies for 10,000 years, if only it was all recoverable! Yet after over 100 years of studying the hydrological cycle, there are no consistent methods to calculate the amount of available and recoverable water from river basins and groundwater systems; few hydrological watershed models even address groundwater in their water balance models (Zeitoun, 2011).

Surface and ground waters that cross international boundaries present increasing challenges to regional stability because hydrologic needs can often be overwhelmed by political considerations. There are 270 international river basins and over 448 transboundary aquifers (Wolf and Giordano, 2002; Bakker, 2009; IGRAC, 2009; Margat and van der Gun, 2013). The basin areas that contribute to international rivers cover approximately 47% of the land surface of the Earth, land on which 40% of the world's population resides; these basins contribute almost 60% of freshwater flow. According to Margat and van der Gun (2013), the permeability architecture of the major transboundary aquifers is not uniform, but rather highly variable, with 35% found in sand and gravel and 18% in complexly deformed (faulted, folded, karst or volcanic) rocks, while 47% are shallow alluvial aquifers or are composed of weathered or fractured rock. The hydrologic significance of this distribution of aquifers is that conflicting conceptual models are common and the use of computer models for decision support is limited to just a small percentage of the aquifers, given that the limiting assumptions for the numerical models are violated for most of the aquifers. Likewise, the significance to groundwater governance is that a catchall approach is not possible, given the highly variable permeability characteristics of the aquifers that store the groundwater.

Environmental flows and ecosystem services are more dependent on groundwater than previously thought. Gautier (2008) estimates that 36% of river runoff comes from groundwater. WaterWorld (2013) reports that new models to map groundwater reveal that ecosystems covering 22% to 32% of Earth's surface rely on groundwater. Yet, most water management and governance schemes suffer from a case of 'hydroschizophrenia' – the creation of separate surface water and groundwater governance and policies, despite the recognition of the hydraulic connection between both hydrologic regimes (M. Llamas, 1975).

A large percentage of the world population also resides in lands underlain by transboundary aquifers. The majority of the world's cities rely on groundwater to some degree for their urban water supplies, and Giordano (2009) posits that developed groundwater contributes to the global urbanization underway today. As a consequence, the global economy is becoming increasingly dependent on groundwater. With groundwater exploitation increasing with growth in population, increasing water needs for food security and the increased awareness of the hydraulic connection of groundwater for environmental flows, the importance of developing ground rules to govern groundwater cannot be overstated.

What is groundwater governance?

The best answer to this question is by Tushaar Shah: 'Groundwater governance worldwide is a work in progress' (Shah, 2009, p205). Water governance, specifically, groundwater governance, means different things to different people. Groundwater governance is like international water resources management as an across-the-board phrase: broad in scope, challenging to evaluate and, given the heterogeneity in both the hydrogeologic settings and socio-ecology of groundwater and aquifer users across the planet, difficult to find one model that might work for all. Given the challenge, Varady et al. (2013) termed this as 'enterprise governance': an opportunity for humans to better understand, through institutions, how groundwater fits into society and its relationship to the planet's environment.

Groundwater governance is variously described in the literature, conference declarations and policy briefs as (1) a process and art form designed to manage groundwater and coordinate decision making (Varady et al., 2013); (2) setting objectives for groundwater, followed by developing strategic institutions and instruments to align stakeholder behavior and outcomes with objectives and local level governance that control outcomes on the ground (Wijnen et al., 2012); and (3) being about creating institutional structures, processes, mechanisms and policy tools for harmonizing the interests of the resource users and the wider interests of society and future generations (Shah, 2009). In short, groundwater governance is about roles, linkages and accountabilities of government institutions and stakeholders, ranging from civil to professional to private actors (Margat and van der Gun, 2013). Defining the concept is the subject of a multiyear, multimillion dollar effort underway by many international organizations such as the World Bank, the Global Environment Facility, the International Association of Hydrogeologists and the United Nations, through UNESCO (United Nations Educational, Scientific and Cultural Organization) and FAO (Food and Agriculture Organization).

As listed in Table 2.1, there are many legal institutions, instruments, regulatory levels and governance options that are used, or have been proposed, for the many models of groundwater governance. The listing in Table 2.1 is far from complete

Table 2.1 Models of groundwater and aquifer governance, policy and management (modified from Nanni et al., 2002; Lopez-Gunn and Jarvis, 2009)

Regulation, policy or management approach	Implications	Limitations	Examples with recent citation where applicable
Minimum legal control	No control over groundwater abstractions.	Reduced natural discharge to ecosystems, pollution.	India, China
Exemption from regulation	Small quantity uses for domestic, stock and industry.	Highly variable quantity of use from state to state and province to province.	United States, Canada (see Chapter 4)
Local customary rules	Groundwater rights defined at local level; mechanisms for local conflict resolution.	Limited controls; no account of impacts to groundwater system, downstream users, water quality.	Pakistan, Iran
Specific groundwater legislation	Well construction and groundwater abstraction controlled.	Little consideration may be given to groundwater dependent ecosystems or water quality.	Philippines

(Continued)

Table 2.1 (Continued)

Regulation, policy or management approach	Implications	Limitations	Examples with recent citation where applicable
Comprehensive water resources legislation	Surface water/ groundwater subject to same regulation; both administered by same agency; water quality regulated under separate agency. Tradable property rights in some regions.	Pollution control may be deficient. Little to no recognition of shallow versus deep groundwater systems.	United States, Canada (Nowlan, 2005; Dellapenna, 2013; Gerlack et al., 2013)
Fully integrated water resources legislation (integrated water resources management)	Integrated catchment/ groundwater body; emphasizes public awareness/ participation; some transboundary issues recognized.	Best chance of implementing balanced and effective regulation policy. Deep aquifers identified if important to ecosystems or drinking water.	European Community (Foster and Ait-Kadi, 2012; Wijnen et al., 2012)
International agreement	Water quality protection, allocation, recharge, extraction.	Surface water/ groundwater interdependence vaguely recognized. Only one agreement in effect for groundwater.	French Prefect de Haute-Savoie and Swiss Canton of Geneva
River basin organization	Management and stakeholder involvement at river basin level.	Marginal recognition of groundwater rights in licensing arrangements.	Murray-Darling Basin, Australia
Groundwater compacts	Proactive in resolving interstate groundwater disputes by interstate compact or some form of agreement.	No examples of two or more states with an interstate compact that solely governs groundwater allocation.	Agreements between U.S. states (Hesser, 2013)
'Crowding out' with supply augmentation	Develop alternative water sources including diversions, aquifer storage and recovery, aquifer storage transfer and recovery, managed recharge.	Expensive. Reexamination of federal laws regarding injected water quality.	Canada, Australia, United Arab Emirates, United States (Nowlan, 2005; Shah, 2009; Henzell, 2012)
	Surface water/ groundwater part of international watercourse.	Best chance of international participation. Not ratified. Deeper 'confined' aquifers not covered.	Convention on the Law of the Non-navigational uses of International Water Courses (McCaffrey, 2007)

Table 2.1 (Continued)

Regulation, policy or management approach	Implications	Limitations	Examples with recent citation where applicable
International conventions	Transboundary aquifers approach with integration of use and water quality protection.	Draft by International Law Commission (ILC) acknowledges importance of conceptual hydrogeologic model.	Draft Convention on the Law of Transboundary Aquifers (ILC, 2008)
	Shallow groundwater connected to surface water. Deeper groundwater part of global commons under 'Law of the Hidden Sea'.	Depth of shallow and deeper groundwater systems based on simple hydrogeologic model.	Adaptation of the UN Commission of the Law of the Hidden Sea proposed by Lopez-Gunn and Jarvis (2009).
Aquifer management organization	Acknowledgment of limited interaction with surface water resources in arid areas.	No recognition of underlying aquifers.	Regional development of Nubian Sandstone in North Africa; Guarani Aquifer in South America
Aquifer authority	Regulatory agency with principal charge of sustaining federally protected aquifer-dependent species.	State legislature can modify withdrawal rates.	Edwards Aquifer Authority, Texas Votteler (2004)
	Aquifer communities designed around private, independent groundwater users and type of aquifer.	Outcome dependent on storage characteristics of aquifers. Quota system designed using swipe card system or other form of payment.	Shah (2009) describes examples from India; similar to Mexico self-monitor community aquifer management. Van Steenburgen (2012) profiles a case study in China.
	Concurrency of aquifer recoverability. Hybridize land use, groundwater recoverability and aquifer storage.	Based on regular well testing and redetermination of aquifer productivity.	United States; county-level implantation in Colorado, Oregon, Utah (Strachan, 2001)
Transdisciplinary subterranean governance	Unitization of aquifer storage. Economic-based approach. Hybridize exploration, storage, extraction and future use of storage. Look beyond aquifers for groundwater.	Based on individual aquifer and redetermination of recoverability. No existing cases. Can be applied to current and future uses of aquifer storage.	Concept applied in Utah; already in use for oil, gas, geothermal, carbon sequestration and hydrofracking (Jarvis, 2011)

(Continued)

Table 2.1 (Continued)

Regulation, policy or management approach	Implications	Limitations	Examples with recent citation where applicable
	Niche diplomacy. Issue or area of expertise where actor concentrates natural and intellectual capital on positive returns on investments recognized by the world.	Niche 'groundwater' and 'aquifer' diplomacy is unfilled opportunity.	Singapore through Singapore International Water Week (Caballero-Anthony and Hangzo, 2012)

and is provided to be more illustrative than comprehensive, but it should be clear that groundwater governance ranges from the individual to international in scope. Aaron Wolf regularly states, 'Water management, by definition, is conflict management' (Wolf, 2008, p51). With so many conflicting models of groundwater governance, it should come as no surprise that the best agreements over groundwater may be to agree to disagree.

Yet the research presented in this chapter suggests that groundwater governance is in many respects in the experimental stage. For my dissertation (Jarvis, 2006), I determined that part of the problem rests with the fact that groundwater was not recognized as part of the hydrologic system when freshwater treaties and other agreements were negotiated; thus, only approximately 15% to 20% of the treaties and other agreements have provisions for groundwater.

Other parts of the groundwater governance problem rest with the fact that the boundaries of the transboundary groundwater systems are very different than the boundaries of a watershed (Richts et al., 2012). Consequently, any treaty or agreement that has a provision for groundwater reflects only a cursory recognition of the groundwater flow system (Matsumoto, 2009).

What is aquifer governance?

Groundwater is often referred to as a classic common pool resource. There appears to be some 'aquifear' among political geographers of connecting the aquifer storing groundwater to the discussion of groundwater governance. This approach is much like water governance managing a river as a 'pipeline', as opposed to the river 'system'. New paradigms of groundwater governance need to acknowledge that the 'common pore resources', the available aquifer storage, must be integrated into governing the subterranean space.

What is an aquifer? A conventional definition found in just about any textbook on groundwater defines an aquifer as 'a hydraulically continuous body of relatively permeable unconsolidated porous sediments or porous or fissured rocks containing groundwater. It is capable of yielding exploitable quantities of groundwater' (Margat and van der Gun, 2013, p255). A definition drummed into my memory during college is that an aquifer is a formation, group of formations or part of a formation that contains sufficient saturated permeable material to yield sufficient, economical quantities of water to wells and springs. Note that the aquifer is the 'container' for groundwater.

But not all definitions of aquifers agree. Consider Article 2 (a) of the Draft Law of Transboundary Aquifers, as developed by the International Law Commission (2008), where an 'aquifer' is defined as 'a permeable water-bearing geological formation underlain by a less permeable layer and the water contained in the saturated zone of the formation'. The significant difference between the technical and legal definitions is the reference to the degree of saturation and the saturated zone. The technical definition acknowledges that an aquifer can be partially saturated, whereas the legal definition only references the saturated zone.

The hydrologic significance between the two definitions focuses on the recharge area of an aquifer. This zone is only partially saturated and, in the eyes of the Law of Transboundary Aquifers, would not be considered an 'aquifer'. Consider Article 2 (g) where the 'recharge zone' is defined as the 'zone that contributes water to an aquifer, consisting of the catchment area of rainfall water and the area where such water flows to an aquifer by runoff on the ground and infiltration through soil'.

Returning to the classical definitions of an aquifer, the principal characteristic of an aquifer is to yield groundwater. Yet, Article 2 (e) of the Law of Transboundary Aquifers defines utilization of transboundary aquifers or aquifer systems as including 'extraction of water, heat and minerals, and storage and disposal of any substance'. Clearly, the new paradigm is that the available aquifer storage is also a shared resource, as described by Puri and Struckmeir (2010). And if one integrates the acknowledgment of troglodytes and stygofauna that live in the saturated and unsaturated portions of an aquifer, such as the Edwards Aquifer in Texas, clearly the discussion about aquifer governance is no longer just about water anymore!

Transdisciplinary subterranean governance

As discussed more fully in the next chapter, the integration of a transdisciplinary approach to groundwater and aquifer governance is also the new paradigm for working across existing disciplines toward new, higher-level synthesis – that is, to 'invent new science'. Transdisciplinary research goes beyond interdisciplinary and multidisciplinary boundaries and paradigms by integrating the practice-oriented character of scientific knowledge with local, practical knowledge and moral concepts and local values (Max-Neef, 2005).

In many respects, synthesis and inventing new science associated with transdisciplinary governance closely coincides with what Kai Lee termed 'civic science' (Lee, 1993, p161). He defined it as 'irreducibly public in the way responsibilities are exercised, intrinsically technical, and open to learning from errors and profiting from success'. The outcomes of a true civic science should be environmental decisions that are at least as good as what would happen otherwise in terms of their (1) conceptual soundness, (2) equity, (3) technical efficiency and (4) practicability.

Aquifer communities

Shah (2009) describes the notion of aquifer communities, where there is recognition that *aquifer conditions* affect human behavior and the institutional response of groundwater users. Shah defines an aquifer community 'as users in a locality who are aware of their interdependence in the development or conservation of a common aquifer or a portion thereof that shapes their individual or collective behavior' (2009, p152). He defined five situations in India that depend on the aquifer characteristics. Note that the governance models in these situations are dependent the storage characteristics of the developed aquifer, not simply on exploring how to share groundwater:

1   Sand and gravel aquifer, high storage and good recharge – no opportunity for aquifer governance because the impact on a typical user is insignificant and no need for aquifer communities. Good governance is planning expansion of groundwater use for maximum social welfare and poverty reduction.
2   Sand and gravel aquifer, high storage with limited to no recharge – no to low opportunity for aquifer governance as long as the users can chase a falling water level and don't recognize the interdependence between groundwater users. Good governance is improved conjunctive management of groundwater, surface water and rainwater.
3   Hard rock aquifer with low storage and some recharge – some opportunity for aquifer governance as the interdependence of users is known leading to a rise in numerous aquifer communities as there is a near zero-sum game of competitive deepening among well owners, but also efficient water use. Good governance is mobilizing water harvesting and decentralized recharge in addition to local demand management.
4   Hard rock aquifer with low storage or alluvial aquifers with confining layer – high opportunity for aquifer community as there is strong recognition of interdependence. Good governance is institutionalizing the positive experience and builds upon them in addition to demand management.
5   Arid or coastal aquifers susceptible to rapidly changing water quality – no opportunity for aquifer governance because the water quantity impact on a typical user is insignificant but the water quality impact is swift. Aquifer communities are incapable of reviving aquifer and exit. Good governance is large-scale public intervention on supply and demand. The goal is to improve management of surface water.

Likewise, van Steenbergen (2012) describes the governance of an aquifer community in Qinxu, one of the counties in Shanxi Province of China. The Qinxu Groundwater Management System regulates all groundwater usage by equipping all 1,473 wells in the county with an operating system that groundwater users operate with individual swipe cards. Integration of online competency within the aquifer community is exemplified by the Digital Water Resource Information Center in the Water Resources Bureau of Qinxu County. The swipe card transactions are transmitted through the Internet to the county center.

The amount of water that can be used is based on a quota that is allocated annually. The quotas vary from area to area and depend on the determination of the sustainable recoverability of groundwater. Quotas are based on the acreage owned, the number of family members and the number of livestock owned. Some wells are very shallow and others are deep, so the volume of water drawn using one unit may vary from 500 to 5,000 liters. Quotas can also be traded between villages and between farmers. According to van Steenbergen (2012) it is more common to see family members and neighbors share 'excess water' than to trade among farmers.

Aquifer communities are part of transdisciplinary subterranean governance by being irreducibly public in the way responsibilities by the stakeholders are exercised. Aquifer communities provide opportunities for process equity and outcome equity. And aquifer communities provide ways to profit from success and to create new identities.

The concurrency experiment or 'prove-it' policies

Concurrency laws are one of the most effective tools for linking water availability and land use at local jurisdictional scales. Since the 1970s, concurrency laws in the United States have typically focused on the availability of public facilities, such as schools, roads, sewers and water supplies, in order to accommodate rapid growth (Strachan, 2001). The state of Arizona was one of the first to do so; in 1973, an Adequate Water Supply Program was initiated. In 1982, Arizona passed the Groundwater Management Act, which creates active management areas where subdivisions lacking 100-year adequate water supplies can move forward if developers disclose the information to the first buyers. Subsequent buyers don't have to be informed.

In 1983, the state of Oregon was one of the first to recognize groundwater limits: based on the low permeability volcanic rocks, Lane County's Code on Subdivisions delineated quantity and quality groundwater limited areas. In 1998, Marion County developed Sensitive Groundwater Overlay Zones Studies due to areas with poor productivity or areas undergoing groundwater depletion in the basalt aquifers.

In 2001, concurrency laws evolved that focused more on exurban growth and groundwater recoverability from aquifers with heterogeneous permeability and associated productivity. For example, in Benton County, Oregon developed an ordinance or 'prove-it' policy for new private developers of subdivisions in 2007, again targeting areas underlain by volcanic rocks with low storage characteristics. The Benton County ordinance was patterned after a comparable ordinance developed in 2007 in Jefferson County, Colorado, where well testing by private landowners was required, due to the construction of subdivisions on fractured igneous and metamorphic rocks with low production and storativity. This phenomenon can best described as an example of 'retrospective predictability', as described by statistician Nassim Taleb in his book *The Black Swan* (2007). Damage to aquifer storage is apparently new to groundwater and aquifer governance; however, the oil industry recognized this dilemma decades ago in developing hydrocarbons from fractured rock reservoirs (Jarvis, 2011).

The fragmented nature of water and land use at the state or provincial level, due in part to the lack of integration between land use and water laws, is leading to a new paradigm in groundwater and aquifer governance that focuses on a 'bottom-up' approach, instead of the traditional 'top-down' approach. Funding shortfalls, the uncertainty associated with the quantitative characteristics of groundwater systems and the growing frustration with the 'dueling expert' situation is leading to increased reliance on a prove-it approach to assertions of adequate water supplies, as well as on periodic retesting of wells for redetermination of water availability, by developers and their consultants alike.

In 2000, Summit County, Utah, began investigating private water systems that were 'running out of water' due to (1) explosive growth associated with meeting the anticipated housing demands for the 2002 Winter Olympics, (2) dwindling yields from area wells and springs in an area dependent on groundwater for 100% of the local supplies and (3) a prolonged drought in the state of Utah. I assisted the Summit County Commissioners with the development of emergency 'concurrency' ordinance limiting the number of building permits issued by the county to the availability of 'wet' water provided by public and private water systems operating in the Snyderville Basin. This instrument came about due to highly variable well yields that were unrelated to groundwater recharge or depletion, but rather were due to damaged and lost aquifer storage.

In all states, the number of homes, condominiums, apartments, hotels, churches, office buildings and the like that can be connected to a public water system was determined by the capacity of their approved sources. In Utah, the Department of Environmental Quality (DEQ) regulates public water systems, rates the capacities of wells and springs and determines the number of connections that can be served. Wells and springs are usually rated by the DEQ at the time of construction or development. The DEQ requires that a new public water supply well be pumped at a constant rate for a minimum of 24 hours. The well is usually rated at 67% (two thirds) of the constant rate used to test the well. Use of this 'two-thirds' rule does not always provide an accurate prediction of yield when a well is pumped on a continuous basis for several days, weeks or months at time.

Like most states, Utah did not require additional testing beyond the initial prove-out testing after well construction. However, Jarvis et al. (2001) revealed that well yields diminished by more than a factor of two in less than 10 years, due to overpumping of the fractured rock aquifers developed in the Snyderville Basin. Conventional groundwater hydraulic theory indicates that each time a water system turns on a well, a portion of the water becomes 'mined' from storage within the aquifer. Few of the Snyderville Basin residents understood that groundwater flows very slowly in the subsurface, and they were often fooled by a good year of precipitation and the concomitant rise in water levels in wells – this did not guarantee that the aquifer had been replenished, however. And most public water systems in the Snyderville Basin did not bother to reevaluate the capacities of their sources until there was a crisis. Given the alarming spread in seasonal aquifer yield in some Snyderville Basin wells, springs and tunnels, prudent water-system design practices dictated sizing service areas to the source capacity observed during the low-yield winter months (see Figure 2.2).

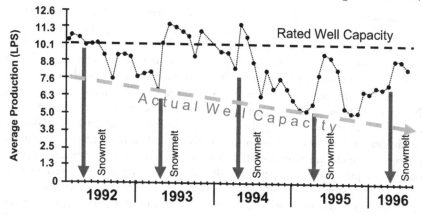

Figure 2.2 Reduced storage capacity in fractured-rock aquifer; modified after Jarvis (2011)

Given the value of land and the economic tie to water in Summit County, some of the water companies considered the water production information to be proprietary. Obvious conflicts arose over how a water system was rated by the Summit County ordinance for current commitments and future growth (Figure 2.2). It was not surprising that attorneys often served as the representatives for some of the water companies, and that traditional litigation was being used in attempts to resolve even the most minor of disputes. After over 10 years of litigation over the constitutionality of the concurrency ordinance and allegations of antitrust, the disputes were settled and an agreement was reached to join the Western Summit County Wholesale Importation Project (Moffitt, 2011). In 2013, the Summit County Council approved an agreement to end the two-decade-long 'water war' that besieged the county (Osowski, 2013).

Concurrency instruments are an adaptive hybrid instrument for groundwater governance that links land use, both past and planned, with groundwater use to aquifer storage. From the perspective of transdisciplinary subterranean governance, the concurrency process for redetermination of aquifer recoverability is technically efficient and irreducibly public in terms of data sharing. Concurrency often limits the rate of private development until wells are crowded out through supply augmentation. The concurrency approach to aquifer governance reflects changing political will, moving beyond 'if we build it, the water will come', to 'if we have it, you are welcome'.

Unitization

Understanding, utilizing and unitizing the underground water resources can ensure water resiliency and security by building aquifer communities. Consider, for example, the governance of nonrenewable groundwater that also recognizes aquifer conditions. Gleick and Palaniappan (2010) defined 'peak' nonrenewable water as groundwater stored in aquifers. The debate over the term 'nonrenewable groundwater' is another source of conflict within the rationale discourse over groundwater and aquifers as summarized by Jarvis (2011), because it refers to groundwater resources where present-day replenishment is limited but aquifer storage is large (Foster and Loucks, 2006), where replenishment is very long (hundreds to thousands of years) relative to the time frame of human use (Foster et al., 2003) or where the use of groundwater storage is at a rate much greater than the renewal rate, essentially 'mining' the groundwater (R. Llamas and Custodio, 2003). Polak et al. (2007) suggest that nonrenewable groundwater resources are essentially 'decoupled' from the hydrologic cycle due to changes in the climatic conditions in the watershed. Narasimhan (2009) indicates that the definition of nonrenewable groundwater also includes aquifers where the storage characteristics of the aquifer have been permanently changed due to pumping, often referred to as 'transient storage'. The storage depletion has been reported to approach 80% of storage space, due to nonelastic deformation in three of the large groundwater artesian basins in the United States: the Dakota sandstone, the Atlantic Coastal Plains and the San Joaquin Valley (Narasimhan, 2009).

Aquifer and oil reservoir storage permanently lost due to the inelastic properties of the sediments and rocks is nonrenewable storage. This is the foundation of the concept of 'unitization' of oil fields that was developed to protect the 'corresponding rights' or 'sovereignty' of all pore space owners in the unit and to not waste valuable pore space.

Unitization is truly a transdisciplinary instrument for groundwater and aquifer governance that acknowledges process equity in both groundwater use and aquifer storage. In the governance and management of oil and gas, 12 countries and 38 U.S. states use unitization. Some states are extending unitization principles to hydrofracking. Unitization is used for the governance of geothermal energy in the Oregon and Utah. The state of Wyoming implements unitization in subsurface sequestration of carbon. Jarvis (2011) describes unitization concepts applied to spiritual connection to springs in Japan. Clyde (2011) and Jarvis (2011) describe the application of unitization concepts to groundwater in the desert regions of Utah.

Unitization is an adaptive hybrid instrument, as it also looks forward to the problem of comparing and contrasting what we think we can develop, using numerical models on our computer screens, to what actually can be developed through well screens through the redetermination process. Unitization, as applied to groundwater and aquifer governance, can lead to (1) promoting groundwater exploration in underutilized areas; (2) preserving the storage capacity of aquifers; (3) promoting private investment in aquifer storage and recovery (ASR), managed recharge (similar to secondary and tertiary recovery operations used in the oil and gas industry), as well as other opportunities, such as remediating contaminated groundwater, ecosystem services and the spirituality of water; and (4) preventing disputes by 'blurring the boundaries' through the creation of aquifer communities. As depicted on Figure 2.3, unitization instruments hybridize exploration, storage and extraction. The process looks 'beyond groundwater', and links past, current and future uses. The redetermination process that is so important for unitization fits transdisciplinary subterranean governance well as it is technically efficient and permits profiting from success.

Stage of Field Life	Unitization Status (Worthington, 2011)	Petroleum Class	Proposed Aquifer Unitization
Exploration		Prospective	Pre-unit Agreement Based on Voluntary Geographic Unit
Discovery		Contingent	
Appraisal	Pre-Unit Agreement		Pre-unit Agreement Based on Voluntary Geologic Unit
Commerciality		Reserves (UD)	
Development	Unitization and Operating Agreement		Unitization and Operating Agreement
Production	Redetermination(s)	Reserves (D)	Redetermination and Compulsory/Conservation Units
Abandonment			Redetermination(s) of New Use of Aquifer

Figure 2.3 Framework for unitization of groundwater

I have received many negative comments from international water lawyers regarding the concept of unitization as applied to aquifers and groundwater. Some of the comments rest with how to integrate 'nature' into the agreement; others focus on how it would be implemented at different scales. The integration of 'nature' or 'ecosystem services' into a unitization agreement would be a simple matter of listing in the unitization agreement and redetermination process, that is, nongovernmental organizations could represent any facet of 'nature' in the agreement. The case study of the connection between spirituality and springs in Japan, as summarized by Jarvis (2011), underscores the plausibility of such a notion. The implementation of a unitization agreement for aquifer governance should not represent any particular institutional challenges, as many countries and states use unitization in the governance of oil and gas already. In my opinion, some of the concerns by the legal industry may spring from fear of competition. For example, a Model Form International Unitization, Unit Operating Agreement and guidelines for the redetermination process were developed by the Association of International Petroleum Negotiators, all of which can be purchased online for a modest cost. We will revisit these immortal words of Hamman (2005, p129) throughout this book: 'Those individuals, communities, and institutions that benefit from the current allocation or perceive they will suffer from a change have great power to defend the status quo.'

Niche diplomacy

Caballero-Anthony and Hangzo (2012) describe niche diplomacy as an issue or area of expertise selected by a country that concentrates its natural and intellectual capital on positive returns on investments recognized by the world. Canada and Norway, with national support for active involvement in international peacekeeping, serve as examples of traditional niche diplomacy.

Singapore secured a niche in water diplomacy that is now recognized across the world as a model for dealing with water scarcity. Singapore developed this niche by capitalizing on innovative water-management and water-treatment technologies, water science and water institutions. One return on Singapore's investment in niche diplomacy is an international conference, the Singapore International Water Week. Singapore now relies on four local sources of water supply, known as the 'Four National Taps': (1) imported water from Malaysia, (2) rainwater harvesting within the local catchment, (3) NEWater – wastewater treated to potable water and (4) desalinated seawater (Caballero-Anthony and Hangzo, 2012).

Other existing examples of niche water diplomacy include the following:

- Scotland is pursuing recognition as the first 'Hydro Nation'. Scottish goals are to (1) reduce energy use, improve efficiency and create a low carbon nation; (2) raise Scotland's international profile through recognition as an international leader on water management and governance; and (3) develop a water center of expertise and research with international reach.
- Sweden has hosted the Stockholm International Water Institute (SIWI) since 1991. SIWI not only performs research, education and training in water governance, transboundary water management, climate change and water, the water-energy-food nexus and water economics, but also organizes the World Water Week in Stockholm, an annual global event on the planet's water issues and related concerns of international development.

The Singapore experience suggests that an opportunity might exist elsewhere in the world for niche water diplomacy in groundwater development and aquifer storage technologies and institutions. Concepts for potential niche water diplomacy as public-private partnerships for groundwater and aquifers include (1) the United Arab Emirates and their use of desalinated water for ASR and (2) Australia and their pioneering work in aquifer storage, transfer and recovery (ASTR).

Many other opportunities exist for niche 'groundwater' diplomacy; it must be considered a key element of future foreign policy. Niche diplomacy rounds out the conceptually sound approach to transdisciplinary subterranean governance by filling the important 'niche' for sharing, from 'what exists' by defining the aquifer community and 'what are we capable of doing' by reducing monopolies increasing competition and accountability (antitrust) through concurrency and the redetermination process, to 'what do we want to do' with the aquifer before, during and after groundwater exploitation (ex ante and ex post), and to 'what must we do' through knowledge sharing and open learning from 'mistakes made' (Figure 2.4).

Even with an international reputation for niche water diplomacy, as part of a multifaceted approach to dealing with water scarcity in Singapore, a new tender was released to explore for groundwater in April 2013 by the national water agency, PUB. This is a surprising development in Singapore's portfolio of water supplies, as groundwater has been historically dismissed as a water supply

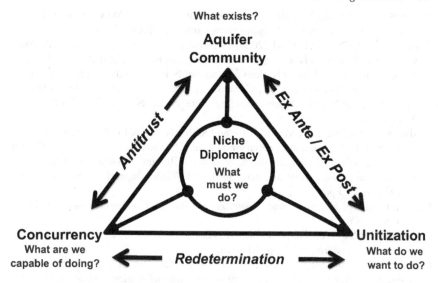

What exists?

Aquifer Community

Antitrust

Ex Ante / Ex Post

Niche Diplomacy
What must we do?

Concurrency
What are we capable of doing?

Redetermination

Unitization
What do we want to do?

Figure 2.4 Framework for transdisciplinary subterranean governance

option. The tender desires to study the potential of tapping aquifers and other underground water sources in the western and southern part of the island, in an area known as the Jurong Formation. Singapore acknowledges the importance of recognizing the connection between the use of both groundwater and aquifers through the following statement: 'the extraction of groundwater will only be carried out if the risks of groundwater extraction can be adequately managed with no impact on existing buildings and infrastructure' (*AsiaOne*, 2013). If use of groundwater is solely dependent on the recoverability of groundwater, then why worry about what happens to the aquifer after the groundwater has been recovered?

References

AsiaOne (2013) 'PUB explores groundwater in western and southern Singapore', 26 April, http://news.asiaone.com/News/Latest+News/Science+and+Tech/Story/A1Story20130426-418568.html

Bakker, M. H. N. (2009) 'Transboundary river floods: Examining countries, international river basins and continents', *Water Policy*, vol 11, pp269–288

Caballero-Anthony, M. and Hangzo, P. K. K. (2012) 'From water insecurity to niche water diplomacy: The Singapore experience', *NTS Insight*, RSIS Centre for Non Traditional Security (NTS) Studies, Singapore, www.rsis.edu.sg/nts/HTML-Newsletter/Insight/NTS-Insight-nov-1201.html

Clyde, S. E. (2011) 'Beneficial use in times of shortage: Respecting historic water rights while encouraging efficient use and conservation', *The Water Report*, vol 83, pp1–13

Dellapenna, J. W. (2013) 'A primer on groundwater law', Villanova University School of Law, Public Law and Legal Theory Working Paper No. 2013-3042, http://ssrn.com/abstract=2265062

Foster, S. and Ait-Kadi, M. (2012) 'Integrated Water Resources Management (IWRM): How does groundwater fit in?' *Hydrogeology Journal*, vol 20, pp415–418

Foster, S. and Loucks, D.P. (eds) (2006) *Non-renewable groundwater resources: A guidebook on socially-sustainable management for water-policy makers*. Paris: IHP-VI, Series on Groundwater No. 10, United Nations Educational, Scientific and Cultural Organization

Foster, S., Nanni, M., Kemper., K., Garduño, H. and Tuinhof, A. (2003) 'Utilization of non-renewable groundwater', Briefing Note Series No. 11, The World Bank, Washington, DC

Gerlak, A.K., Medgal, S.B., Varady, R.G. and Richards, H. (2013) 'Groundwater governance in the U.S. – summary of initial survey results', Summary report by the Udall Center for Studies in Public Policy and the Arizona Water Resources Research Center, https://wrrc.arizona.edu/sites/wrrc.arizona.edu/files/pdfs/GroundwaterGovernanceReport-FINALMay2013.pdf

Gautier, C. (2008) *Oil, water, and climate: An introduction*, Cambridge University Press, New York, NY

Giordano, M. (2009) 'Global groundwater? Issues and solutions', *Annual Review of Environment and Resources*, vol 34, pp7.1–7.26

Gleick, P.H. and Palaniappan, M. (2010) 'Peak water: Conceptual and practical limits to freshwater withdrawal and use', *Proceedings of the National Academy of Sciences (PNAS)*, vol 107, no 25, pp11155–11162, www.pacinst.org/press_center/press_releases/peak_water_pnas.pdf

Hamman, R. (2005) 'The power of the status quo', in J.T. Scholz and B. Stiftel (eds) *Adaptive Governance and Water Conflict: New Institutions for Collaborative Planning*, Resources for the Future, Washington, DC, pp125–129

Henzell, J. (2012) 'Ensuring the security of water, "a strategic commodity on par with oil"', *The National*, 30 June, www.thenational.ae/news/uae-news/environment/ensuring-the-security-of-water-a-strategic-commodity-on-par-with-oil

Hesser, J.N. (2013) 'The nature of interstate groundwater resources and the need for states to effectively manage the resource through interstate compacts', *Wyoming Law Review*, vol 11, pp25–46

ILC-International Law Commission (2008) 'Official records of the General Assembly, Sixty-third Session, Supplement No. 10', (A/63/10). Draft articles on the Law of Transboundary Aquifers, http://untreaty.un.org/ilc/texts/instruments/english/draft%20articles/8_5_2008.pdf

International Groundwater Resources Assessment Centre (IGRAC) (2009) 'Transboundary aquifers of the world', www.igrac.net/publications/320, accessed September 2009

Jarvis, W.T. (2006) 'Transboundary groundwater: Geopolitical consequences, commons sense, and the Law of the Hidden Sea', unpublished PhD dissertation, Oregon State University, Corvallis, OR, http://hdl.handle.net/1957/3122

Jarvis, W.T. (2011) 'Unitization: A lesson in collective action from the oil industry for aquifer governance', *Water International*, vol 36, no 5, pp619–630

Jarvis, W.T. (2013) 'Water scarcity: Moving beyond indices to innovative institutions', *Groundwater*, vol 51, no 5, pp663–669

Jarvis, W.T., Yonkee, A. and Matyjasik, M. (2001) 'Transient storage and compressibility of fractured bedrock aquifers, implications for aquifer sustainability, Twin Creek Limestone, Utah: Proceedings of 2001 annual meeting', American Institute of Hydrology.

Lee, K.N. (1993) *Compass and gyroscope*, Island Press, Washington, DC

Llamas, M. (1975) 'Non-economic motivations in ground water use: Hydroschizophrenia', *Groundwater*, vol 13, pp296–300

Llamas, R. and Custodio, E. (2003) 'Intensive use of groundwater: A new situation which demands proactive action', in R. Llamas and E. Custodio (eds) *Intensive Use of Groundwater: Challenges and Opportunities*, Balkema, Lissa, the Netherlands, pp13–31

Lopez-Gunn, E. and Jarvis, W.T. (2009) 'Groundwater governance and the Law of the Hidden Sea', *Water Policy*, vol 11, pp742–762

Margat, J. and van der Gun, J. (2013) *Groundwater around the world: A geographic synopsis*, CRC Press/Balkema, Leiden, the Netherlands

Matsumoto, K. (2009) 'Appendix E: Treaties with groundwater provisions', in J. Delli Priscoli and A.T. Wolf (eds) *Managing and Transforming Water Conflicts*, Cambridge University Press, New York, NY, pp266–273

Max-Neef, M.A. (2005) 'Foundations of transdisciplinarity', *Ecological Economics*, vol 53, pp5–16

McCaffrey, S.C. (2007) *The law of international watercourses, non-navigational uses*, Oxford University Press, New York, NY

Moffitt, S. (2011) 'Anti-trust settlement between courthouse, Summit Water, reached', *The Park Record*, 11 November

Narasimhan, T.N. (2009) 'Groundwater: From mystery to management', *Environmental Research Letters*, vol 4, no 3, doi:10.1088/1748–9326/4/3/035002

Nanni, M., Foster, S., Dumars, C., Garduño, H., Kemper, K. and Tuinhof, A. (2002) 'Groundwater legislation and regulatory provision. Groundwater dimensions of national water resource and river basin planning. Sustainable groundwater management concepts & tools', Briefing Note Series No. 4, The World Bank, Washington, DC

Nowlan, L. (2005) 'Buried treasure: Groundwater permitting and pricing in Canada', with Case Studies by Geological Survey of Canada, West Coast Environmental Law and Sierra Legal Defense Fund, www.waterlution.org/sites/default/files/gw_permitting_pricing_canada_e.pdf

Osowski, A. (2013) 'County approves water agreement', *The Park Record*, 21 June, www.parkrecord.com/news/ci_23513909/county-approves-water-agreement

Polak, M., Klingbeil, R. and Struckmeier, W. (eds) (2007) *Strategies for the sustainable management of non-renewable ground water resources*, Federal Institute for Geosciences and Natural Resources (BGR), Hannover, Germany

Puri, S. and Struckmeier, W. (2010) 'Aquifer resources in a transboundary context: A hidden resource? Enabling the practitioner to 'see it and bank it' for good use', in A. Earle, A. Jägerskog and J. Öjendal (eds) *Transboundary Water Management: Principles and Practice*, Earthscan, London, UK, pp73–90

Richts, A., Struckmeier, W. and Foster, S. (2012) 'WHYMAP: Global map of river and groundwater basins at the scale of 1:50,000,000', World-wide Hydrogeological Mapping and Assessment Programme, BGR, Germany

Shah, T. (2009) *Taming the anarchy: Groundwater governance in South Asia*, Resources for the Future, Washington, DC

Strachan, A. (2001) 'Concurrency laws: Water as a land-use regulation', *Journal of Land, Resources, and Environmental Law*, vol 21, p435

Taleb, N.N. (2007) *The black swan: The impact of the highly improbable*, Random House, New York, NY

van Steenbergen, F. (2012) 'A brave new groundwater world', *TheWaterBlog*, 22 October, www.thewaterchannel.tv/en/thewaterplaza/the-water-blog/123-a-brave-new-groundwater-world

Varady, R.G., van Weert, F., Megdal, S.B., Gerlak, A., Abdalla Iskandar, C. and House-Peters, L. (2013) 'Groundwater policy and governance', Thematic Paper No. 5,

Commissioned by UNESCO IHP, Groundwater Governance: A Global Framework for Country Action

Votteler, T. H. (2004) 'Raiders of the lost aquifer? Or, the beginning of the end to fifty years of conflict over the Texas Edwards Aquifer', *Tulane Environmental Law Journal*, vol 15, pp258–335

WaterWorld (2013) 'Groundwater depth mapped but data still lacking from developing countries', 17 May, www.waterworld.com/articles/2013/05/groundwater-depth-mapped-but-data-lacking.html

Wijnen, M., Augeard, B., Hiller, B., Ward, C. and Huntjens, P. (2012) 'Managing the invisible: Understanding and improving groundwater governance', http://water.worldbank.org/publications/managing-invisible-understanding-and-improving-groundwater-governance

Wolf, A. T. (2008) 'Healing the enlightenment rift: Rationality, spirituality and shared waters', *Journal of International Affairs*, vol 61, no 2, pp51–73

Wolf, A. T. and Giordano, M. A. (2002) *Atlas of international freshwater agreements*, United Nations Environment Programme, Nairobe, Kenya

Zeitoun, M. (2011) 'The global web of national water security', *Global Policy*, vol 2, no 3, pp286–296

3 Water negotiation frameworks

This chapter will introduce the reader to the fields of negotiated and participatory approaches for water security and related risks, water allocations and related interests, and transforming conflicts over a broad spectrum of water issues. These topics have been an active area of scholarship and practice for nearly 40 years. I have not found a comprehensive bibliography of these fields, but scholarly articles and the gray literature must surely number in the thousands (see Delli Priscoli and Wolf, 2009, for one of the best listings). The literature spans many different domains and experiences from across the globe. The academic disciplines that frequently publish in this arena are hydropolitics, planning, geography, conflict resolution and law. With the advent of computers to simulate different hydrologic development scenarios, we are starting to see engineers and hydrologists increasingly wading into the game.

It is clear that these fields overlap to a great extent, and that issues in one area of interest, for example, national water security, may also have connections to programmatic interests, for example fisheries, which may also have connections to individuals whose identity is closely, if not completely, connected to fishing as their livelihood. As such, addressing such complex and diverse topical issues indicates the solution to water negotiations may benefit greatly by exploring negotiations through a transdisciplinary imagination. A new negotiation framework designed around transdisciplinarity is offered toward the end of this chapter.

What are the best approaches to negotiations over water? The answer to this question mimics the problem of defining the concepts of safe yield and sustainability in the hydrogeology discussed in earlier chapters – the best approach depends on whom you ask and when you ask. Negotiation frameworks come with many names and forms. I spent a year traveling to the locations hosting training in the major water negotiation frameworks in order to learn the intricacies beyond what one could decipher from the literature. What I found is that few of the water negotiation frameworks acknowledge the existence of the other frameworks. However, common traits among the different frameworks include case studies, role plays and scenarios. The following is a short summary of water negotiation frameworks found in the literature.

Loops, knots, webs and nets

The ferryman Charon from Greek mythology more than likely used a variety of tools to guide his crazy boat across the River Styx. Ropes tied into a variety of loops, knots, webs and nets, along with a variety of devices for orienteering and navigation, are needed by skilled captains to guide passage and safeguard passengers. Boat operators have used tools to splice, repair and weave 'new' tools. Navigation skills and tools are often used as allegories for conflict resolution and negotiations. I use analogies to navigation tools because they are useful in developing mental models of the many water negotiation frameworks.

Loops

Circles, or 'loops', are a common form used in mental models of conflict resolution. Argyris and Schön (1996) provide some of the first mental models using loops and double loops in learning networks. Daniels and Walker (2001) use a loop as a framework for conflict management in moving from assessment to strategy to implementation. According to Innes and Booher (2010), their collaborative rationality framework is similar to the 'model II theory' as defined by Argyris and Schön (1996), where interaction is based on obtaining valid information, making informed choices and assuring internal commitment to the choices.

Intersecting loops are a common method of depicting the interaction or meshing of concepts. Although Abukhater (2013) does not use a mental model to describe his 'process equity–outcome equity' framework, it is built around three key components: rules of engagement, mechanisms of engagement and neutral third-party mediation. In the process equity loop, procedural justice is ensured through the negotiation process, planning analysis and structure of agreements, all of which are overprinted by context or control factors that he refers to as 'cultural hydrology'. For the outcome equity loop, distributive justice is assured by the equity parameters that are both commensurable and perceived, or that move from 'rights to needs' to 'needs to satiety'.

The Water Diplomacy Framework of Islam and Susskind (2013) utilizes the chain or interaction of natural and societal domain variables within the political domain as shown in Figure 3.1. Within the natural domain, they propose that the

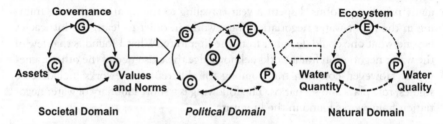

Figure 3.1 Water Diplomacy Negotiation Framework showing interactions among natural and societal processes within a political domain; from Islam and Susskind (2013)

triple constraints on water – quantity (Q), quality (P) and ecosystem (E) – and their interdependencies and feedback may lead to constraints and conflicts that cannot be separated from societal domain variables – social values and norms (V), economy (C) and governance (G).

While they acknowledge that the notion of linkages between the natural and societal domains is not unique to their framework, Islam and Susskind (2013) argue that boundary spanning between these two domains is necessary to address wicked water problems. There are six fundamental elements that are part of or influence the Water Diplomacy Framework:

1   Water that crosses domain (natural, societal, political) and boundaries at different scales (space, time, jurisdictional, institutional)
2   Embedded water, blue and green water, virtual water, technology sharing and negotiated problem solving to arrange for reuse can 'create flexibility' in water for competing demands
3   Water networks are made up of societal and natural elements that are open and cross boundaries and that change constantly in unpredictable ways within a political system
4   All stakeholders need to be involved at every decision-making step, including problem framing and formulation, investments in experimentation and monitoring are key to adaptive management, and the process of collaborative problem solving needs to be professionally facilitated
5   Stakeholder assessment, joint fact finding, scenario planning and mediated problem solving
6   The Mutual Gains Approach (MGA) to value creation; multiparty negotiation keyed to coalitional behavior; mediation as informal problem solving

Key questions that are addressed when implementing the Water Diplomacy Framework include the following:

1   Have all the right parties been included adequately?
2   Was joint fact finding or other fact-finding undertaken?
3   Have there been occasions for profession-neutral facilitation of joint problem solving during which 'inventing without committing' has been encouraged?
4   Has there been a search for nonzero sum options or packages that link issues creatively or build upon possible technology innovation?
5   What consideration has been given to collaborative adaptive management? And, what efforts have the parties made to review and adjust the solution or decision over time in light of changing conditions?
6   What effort has been made to encourage institutional capacity building or organizational learning?
7   Did the decision(s) or solution(s) presented in this case have political credibility, and if so, why?
8   How was the decision/solution implemented?
9   What metrics were used to measure the effectiveness of solutions and/or decisions?

Following my review of Islam and Susskind (2013; see Jarvis, 2012) coupled with attending a workshop on the Water Diplomacy Framework at Tufts University, I left many of the lectures and the daily role plays with the conclusion that this framework drove home the point that nonzero sum or interest-based negotiations approaches to water negotiations are fundamental in this framework.

Knots

The 'knot' or 'trifecta' form is another common form used as a mental model for collaboration or sustainability. A 'trifecta' occurs when three elements come together at the same time, sometimes represented by a simplified triquetra symbol or three-sided knot representing a triangle or the intersection of three domains. The trifecta is attributed to the Celts where it represents the interconnection of three planes of existence, for example, the mind, body and spirit. We see trifectas in many fields related to natural resources. Perhaps the most famous 'trifecta' is the Venn diagram used to portray the notion of the triple bottom line model of sustainability. The trifecta commonly appears in diagrams used to portray cooperation in the field of conflict resolution. The most famous example is the logo for the online resource Mediate.com (Figure 3.2).

A framework focusing on international water law is introduced by Paisley (2008), who argues for the need to integrate water negotiations with international water law. His argument is based on the premise that transboundary water agreements are an important part of building confidence and trust between states regardless of scale. Why a special framework dedicated to international water law? Paisley (2008) proffers that the rules of international law are the building blocks in managing and maintaining relations between entities. Sovereignty is the driving force for actions and activities within state boundaries. These activities have boundaries, as these activities may impact neighboring states as well as distant states.

According to Paisley (2008, p15), 'International law encompasses global, multilateral or bilateral agreements, as well as customary law, state practice, institutions that develop and administer the law and the extra-territorial application of domestic law. Among other things, international law attempts to control, limit and prevent environmental damage and promote a clean and healthy environment. Environment is a broad topic, including fresh and salt water, soil, land, atmosphere, all living creatures and all other aspects of the physical environment.'

As depicted in Figure 3.3, international law comes in primarily three forms. Conventions, treaties, agreements, protocols, and so on, that are mutually agreed upon written instruments between states constitute 'hard law'. When it comes to

Figure 3.2 Mediate.com knot-based logo

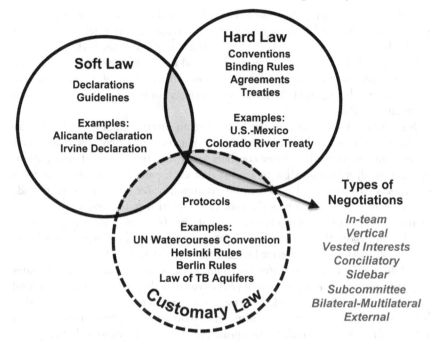

Soft Law

Declarations
Guidelines

Examples:
Alicante Declaration
Irvine Declaration

Hard Law
Conventions
Binding Rules
Agreements
Treaties

Examples:
U.S.-Mexico
Colorado River Treaty

Protocols

Examples:
UN Watercourses Convention
Helsinki Rules
Berlin Rules
Law of TB Aquifers

Customary Law

Types of Negotiations

In-team
Vertical
Vested Interests
Conciliatory
Sidebar
Subcommittee
Bilateral-Multilateral
External

Figure 3.3 International water law negotiation framework

international water law, there are hundreds of treaties describing binding rules on the sharing of rivers; most of these can be acquired through the Transboundary Freshwater Dispute Database maintained by Oregon State University (Oregon State University, 2013). Only a few treaties and agreements exist for shared groundwater resources (Matsumoto, 2009; International Water Law Project, 2013).

'Soft law' refers to documents that are not legally binding, such as declarations, guidelines, resolutions and statements of principle or codes of conduct. The many United Nations resolutions and statements from major UN bodies are good examples of soft law. It has become commonplace for water conferences, symposia and the like to develop declarations such as the Alicante Declaration and the Irvine Declaration, among many others that proclaim support for different facets of water availability, rights, protection, sanitation, economics and so on. Paisley (2008) indicates that soft law is important because it many times serves as the foundation for later, legally binding agreements.

Customary international law applies if states sharing international freshwater resources are not parties to a treaty or agreement. Customary law and the associated rules can be interpreted in many different ways; each state may have differences of opinion regarding the rules. However, the codification of customary law removes some of these ambiguities. The United Nations Convention on the Law of the Non-navigational Uses of International Watercourses (1997) is one of the most comprehensive international water law treaties concluded under United Nations auspices (McCaffrey, 2007). Paisley (2008) indicates that the

UN convention is generally regarded as reflecting the fundamental rules of customary international law applicable in the international water field.

The key principles under the international water law framework include the following: (1) the duty not to cause substantial injury; (2) the right to an equitable and reasonable share in the utilization of the waters of an international watercourse or basin; and (3) the duty to inform, consult and engage in good faith negotiations (Paisley, 2008).

Negotiation is important in the context of an international water law framework. Negotiation and transboundary water agreements help countries move away from the view that water conflicts are a zero-sum game. If negotiations are successful, it is reasonable to assume that each party will benefit. The types of negotiations evaluated under an international water law framework include (1) horizontal or in-team negotiations, (2) vertical negotiations with superiors or constituents, (3) vested interest negotiations, (4) conciliatory negotiations, (5) sidebar negotiations, (6) subcommittee negotiations, (7) bilateral or multilateral negotiations and (8) external negotiations (Paisley, 2008).

While an international water law framework is an important part of water negotiations, Allan and Mirumachi (2010) caution that 'those who expect legal principles to be a sound basis for sharing international waters reach for tools that have proved to be wanting as a principled basis for the achievement of water security. The sound and enlightened principles such as *reasonable and equitable* use as well as the awkward ones shaped by *sovereignty or prior use* have proved to be inadequate. Often these apparently obvious ways of engaging with the problem of sharing water resources play a minor role in the actual achievement of strategic water security'.

Webs

A web is a symbol of a complex, yet ordered, interconnected structure or arrangement. A web can also connote something intricately contrived or something that ensnares or entangles. The Water Security Framework is an important part of the negotiator's toolbox, since many water negotiations hinge upon this paradigm. Yet, there is no consistent definition of water security. Nor is there a way to measure water security, even if it could be defined with a reasonable degree of certainty.

Zeitoun (2011) promotes the 'web' approach to water security, as depicted in Figure 3.4. The web provides a good mental model for a Water Security Framework that is broad, interdisciplinary in analysis and cross-sectoral in application. The web as a conceptual tool emphasizes the inseparability of social and biophysical processes related to water and an understanding of how these mediate and are mediated by the socioeconomic and political context within which they occur. Rather than a summary of the paradigm as viewed by Zeitoun (2011), consider the following quote:

> The 'web' of water security identifies the 'security areas' related to national water security. These include the intimately associated natural 'security resources' (water resources, energy, climate, food), as well as the security of

the social groups concerned (individual, community, nation). The 'web' recognizes the interaction occurring at all spatial scales, from the individual through to river basin and global levels. In this sense, an individual's water security may coexist with national water insecurity.

(Zeitoun, 2011, p290)

A related definition was developed by Wouters (2013, p1), who suggests that water security is '[t]he capacity of a population to safeguard sustainable access to adequate quantities of and acceptable quality water for sustaining livelihoods, human well-being, and socio-economic development, for ensuring protection against water-borne pollution and water-related disasters, and for preserving ecosystems in a climate of peace and political stability'.

Tacitly implied in both Zeitoun's (2011) and Wouters's (2013) definitions is the notion of risk – risks associated with floods, droughts and pollution. What is missing from the definition is the risk associated with terrorism.

Conversely, Tindall and Campbell (2012, p1) define water security as 'the protection of adequate water supplies for food, fiber, industrial, and residential needs for expanding populations, which requires maximizing water-use efficiency,

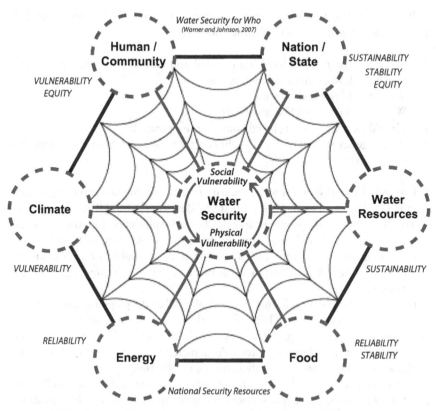

Figure 3.4 The global 'web' of national water security; from Zeitoun (2011)

developing new supplies, and protecting water reserves in the event of scarcity due to natural, man-made, or technological hazards. Eco-system function, environmental, social, and economic parameters are composite water-security components . . . water security and other security measures should be thought of as sustainability, not merely physical elements – there can be no security without sustainability'.

These specialists indicate that water security should be based on an 'all-hazards' approach that includes disruption from the primary hazards – anthropogenic (terrorist or manmade), natural and technological. For example, the U.S. Department of Homeland Security (2013) addresses the water and wastewater sector as 'vulnerable to a variety of attacks, including contamination with deadly agents, physical attacks such as the release of toxic gaseous chemicals and cyber attacks'.

The questions that are addressed in a Water Security Framework focus not only on the fundamental questions associated with hydropolitics, as outlined by Lasswell (1936) – that is, who gets what, when, where and why – but also the following:

1   Water security for whom?
2   Who gets left out?
3   What about water quality as well as quantity? (see Cascão and Zeitoun, 2010)

If one adds the notion of water security posited by Tindall and Campbell (2012), the list grows to include these questions:

4   What are the roles of local, state, tribal and federal governments in water resources development and management?
5   Who should pay and how much?
6   What agencies should be involved?
7   Should existing projects be revamped or re-operated?
8   What agency should have oversight control for security from a sustainability perspective?

I attended a workshop on water security frameworks at the University of East Anglia (see Lankford et al. 2014 for material developed from UEA courses) and, given all of the many questions and the lack of well-defined metrics to assess water security, I left with my own definition: 'water security is water use and reuse without getting into trouble'. The definitions of trouble for each situation, or the acceptable risk for given investments, can then be the starting points for negotiation.

In a related 'web'-based format, Dore et al. (2010) developed the 4Rs Framework for water negotiations. Their framework is based on the World Commission on Dams concepts of 'rights and risks', and they added 'responsibilities and rewards' in their negotiations analysis. As depicted in Figure 3.5, the 4Rs Framework focuses on rewards, risks, rights and responsibilities.

Rewards come in many forms in the 4Rs Framework, ranging from benefit sharing to new regimes, ecosystem services and other incentives. The key questions

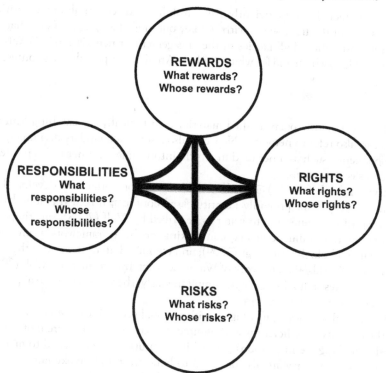

Figure 3.5 Four Rs Negotiation Framework; modified after Dore et al. (2010)

in this part of the framework include (1) What are the possible rewards for the different options? (2) Who are the winners and losers from the rewards? (3) How are rewards shared? And (4) What is fair, effective and sustainable?

Risks are commonplace in water use, management and development; they extend beyond the financial sector and they also include voluntary and involuntary risks. Dore et al. (2010) argue that risk in the 4R Framework should focus on involuntary risks. The key questions to address in this part of the framework are (1) Who are the voluntary risk takers? (2) Who are the involuntary risk bearers? (3) How might risks be shared? And, especially, (4) How might involuntary risks be reduced?

The traditional notion of rights through the lens of legal standing is supplemented by including the position of the human right to water or a water-related service in this part of the 4Rs Framework. Dore et al. (2010) acknowledge the issue of rights is a complex issue; key questions in this part of the framework include (1) What are the rights of all parties in the negotiation? (2) Are there overlapping rights? And (3) What are the different views on prioritization of rights?

Dore et al. (2010) argue that all stakeholders in water have responsibilities. Access to water from a river or captured from an aquifer entails responsibility to use the extracted water efficiently. Responsibility and accountability are critical to this part of the framework, with the key questions focusing on (1) What are the responsibilities of all parties in the engagement or negotiation? (2) Who is accountable to whom and for what? And (3) Are these responsibilities contested?

Nets

Nets are commonly open-meshed material used typically for catching 'things'. Nets can also refer to networks. Nets can also refer to a boundary dividing space for opponents, such as a net used in a badminton game, or a net to keep insects out of a sleeping area.

Jarvis and Wolf (2010) used the ARIA framework originally developed by Rothman (1997) for primarily identity-based conflicts; linked this foundation to the spiritual connection to water as described by Wolf (2008); superimposed the core motives influencing decision making, or the 4i framework as proposed by social psychologist van Vugt (2009); and combined all of these into the 'net' of the Four Worlds/4i Framework of Water Conflict Transformation. Wolf (2008, 2012) indicates that all stages exist simultaneously; there is no need to approach the 'net' in sequence for success.

Adversarial settings typify Stage I, with regional geopolitics often overwhelming the capacity for efficient water resources management. The core motives for decision making are institutions, and while institutions are designed to manage limited resources, they are also designed to build trust and invoke fairness. Key questions to ask at this stage are, What was and what is the issue?

As the adversarial stage plays out in Stage II, the strict rights-based, country or state/province-based positions of each side may play out, but in reality the actual

Negotiation Stage	Common Water Claims	Collaborative Skills	Geographic Scope	Core Motive Influencing Decision Making
Adversarial	Rights	Trust Building	Nations	Institutions
Reflexive	Needs	Skills and Competency Building	Watersheds	Information
Integrative	Benefits	Consensus Building	Benefitsheds	Incentives
Action	Equity	Capacity Building	Region	Identity

Figure 3.6 Four Worlds/4i Negotiation Framework; modified after Jarvis and Wolf (2010)

water negotiations during this stage can last decades. In the 4i framework, information gathering promotes listening by each party and joint fact finding. Van Vugt (2009) indicates that with better information, parties face less uncertainty and can move toward making more sustainable choices. A key question to ask stakeholders is, What could be?

In Stage III participants have moved from thinking about rights to thinking about needs, the problem-solving capabilities that are inherent to most groups can begin to foster creative, cooperative solutions. Van Vugt (2009) indicates that incentives are the motivators for appealing to people's desires to enhancing themselves through seeking pleasure and avoiding pain. A key consideration at this point is, What is the underlying the desire for the resource?

Finally, Stage IV helps with tools to guide the sustainable implementation of the plans and to ensure that the benefits are distributed equitably among the parties. The scale at this stage is now regional, where political boundaries are replaced, thus reintroducing the political interest in seeing that the baskets that have been developed are to the benefit of all. The collaborative learning emphasis is on capacity building, primarily of institutions. Under the 4i framework van Vugt (2009) indicates that identity works toward action by connecting groups of competitors to move toward action. He indicates it is important to create superordinate identities such as regions by thinking of ways to 'blur group boundaries'. Key questions addressed at this stage include (1) How can benefits be distributed equitably or perceived as fair? (2) How can sustainable and resilient institutions be crafted? And (3) How are the existing institutions and organizations to be taken care of or compensated for any change?

I teach a workshop with Aaron Wolf at Oregon State University using the Four Worlds/4i Framework. Given the foundation of the framework around Rothman's (1997) work on identity-based conflicts, coupled with Wolf's (2008, 2012) work on connecting spirituality to water negotiations, then integrating van Vugt's (2009) 4i framework on the social psychology of decision making, ultimately leading to a new identity, this framework focuses on turning inward before looking outward, or the power of identity.

Even new modalities of negotiations organize around 'nets'. Daniels et al. (2012) describe their Unifying Negotiation Framework for climate change negotiations around the ability to juggle three different levels at the same time:

1 Managing the cultural and institutional context of the situation – that is, power structures, norms, practices, history, legislation and rules around the process – in order to create a process with a realistic likelihood of generating an implementable agreement.
2 Organizing the process itself – for example, structuring workshops, designing activities, establishing dialogue and steering the negotiation among the stakeholders.
3 Managing whatever is in the room – the people, issues, history, emotions, concerns, worries, claims, blame – and whatever flows out from the participants in the moment.

When one examines the negotiation frameworks around water, what is missing from each is the acknowledgment of different competencies, different scales and different programs. Nearly all focused at the international scale, some focused on high-level programs (water ministers or high-level water managers) and most focused on surface water, and the focus of professional competencies varied widely, with most emphasizing the more integrative knowledge/cognitive competencies in combination with a strong focus on personal and value/ethical competencies. A few acknowledged the importance of cultural competency or different negotiation styles. What was missing was an acknowledgment that all of the negotiation frameworks were important to resolving debates over investments, risk, interests and identity; that all program skills needed to be integrated into the negotiations; that all competencies were important; and that a single water conflict more than likely rippled through many scales ranging from intrapersonal, to expert-to-expert, to watershed and aquifer, to transboundary rivers and aquifers, to different sectors (energy, water, food), to many agencies (irrigation, water rights, fisheries, etc.) and to many different cultures and many states. Water conflict resolution and negotiation over the use and reuse of water is inherently transdisciplinary.

The Hydro-Trifecta Framework

The Hydro-Trifecta is a new framework that represents the synthesis of the three principal water conflict transformation and negotiation frameworks as they are integrally related. It is like a 'fid', the conical tool made from wood or bone, used to hold open strands of rope for splicing. The Hydro-Trifecta Framework weaves all of the strands of the water negotiation frameworks together. When the water conflict and negotiation frameworks are woven together, they form a 'trifecta'.

The philosophy of the Hydro-Trifecta Framework is to help people navigate through the many issues that are directly related to negotiations over groundwater and aquifer resources. As depicted in Figure 3.7, the framework is designed around the trifecta 'knot' that serves as the 'sighting mechanism' of a 'compass'.

Geologists and hydrologists have used compasses or pocket transits since the late 1800s to measure azimuth angles or bearings, as well as many things associated with their work – the orientation, or strike and dip, of the strata that serve as aquifers; the attitude of fractures and folds that influence groundwater circulation; the grades and slopes important for assessing how water or landslides might move; and the inclination of veins, both seen, such as a vein filled with calcite or ore, and unseen, such as those reported to geologists by water dowsers.

Compasses generally have four main parts: (1) the case or housing that can have many devices for measurements, but most important is a bubble level; (2) the orienting arrow or 'sight' for taking measurements; (3) a compass needle, or floating card, that is balanced on a pin and acts as a pointer; and (4) a graduated circle to determine compass bearings. The Hydro-Trifecta Framework operates somewhat like a compass, with targeted skills serving as the case and baseplate of the compass to take measurements and 'level' the playing field in negotiations. The trifecta knot serves like the needle and sighting device, orienting

Figure 3.7 Hydro-Trifecta Framework

and guiding the direction of the negotiations by using the three frameworks of water security, water diplomacy and water conflict transformation. And the graduated circle provides the transdisciplinary directions for questions that must be addressed in water negotiations – from 'what exists', incorporating fields of study such as geology and economics, to 'what we are capable of doing', focusing on the fields of engineering and commerce, and from 'what we want to do' through planning, law and policy, to 'what we must do', based on values, ethics and philosophy (Max-Neef, 2005).

The compass helps us think of water negotiations not only as a journey but also as a way to collect data. These data can in turn be used to help guide our decisions when we come to a fork and must choose a path to follow. The notion of a compass used in this way calls to mind the statement by Rittel and Webber (1973, p162, emphasis added) on wicked problems: 'One cannot understand the problem without knowing about its context; one cannot meaningfully search for information without the *orientation* of a solution concept; one cannot first understand, then solve'.

The use of maps and navigation devices as allegories to various aspects of conflict resolution and negotiation is not new. Lee (1993) referred to sustainability as the 'map' on which the connections between science and politics are played out. His 'compass' connects to adaptive management. And conflict resolution is Lee's 'gyroscope' which stabilizes the 'boat' in rough waters. What sets the Hydro-Trifecta Framework apart from Lee's seminal work is that it builds upon, rather than duplicates, the use of these 'instruments' to navigate conflicts over groundwater and aquifers through the transdisciplinary imagination.

Building the compass – the case of skills

Scale targeted skills

Scale has many dimensions and definitions. Some practitioners argue that the best scale for water issues should focus on the scales inherent in integrated water resources management (IWRM) because of its wide acceptance and because it recognizes that problems can range from the professional-to-professional level to project team, local government, regional, national and international levels (Kennedy et al., 2009; Foster and Ait-Kadi, 2012; Wolf, 2012). Shah (2009) argues that IWRM does not work well for aquifers and indicates that the problem with IWRM and the global groundwater governance debate is that it transforms, all at once, a predominantly informal water economy into a formal one.

In tackling wicked problems, Dovers (2010) suggests that it is useful to go beyond the scale that is understood and influences the approach of a discipline or subdiscipline and instead explore why that particular scale is used. The goal is to examine the apparent scale versus the underlying logic or the 'embedded scale'. Collaboration between disciplines and local knowledge systems requires understanding that the underlying logic of the scale adopted or implied may be just as important as knowing what that scale is. Dovers (2010) recommends mapping each context, as each situation will more than likely vary across specific settings and issues even in a single jurisdiction.

The Hydro-Trifecta Framework uses conflict as the scaling mechanism because it can be mapped, and the 'embedded scale' can be reasonably assessed. For example, at the intrapersonal scale, the embedded scale may be consumption. At the interpersonal scale, the embedded scale may be professional reputation. At the intersectoral and interagency scales, the embedded scale may be agreements. At the interstate and international scales, the embedded scale may be compacts or treaties. To complicate matters, scales evident in the situation for one societal concern (e.g., well owners) will be different than those for other concerns (e.g., state agencies) and will include multiple issues and multiple interactions between a number of institutional-scale complexes. And yet, these scales can be mapped using systems thinking and situation maps.

Competency targeted skills

Lifelong learning is key to developing functional and personal competencies during education, through work and within the profession as an individual moves up the 'slope' from junior level professionals to managers or ministers. Competency increases confidence. When a person is confident, he or she feels less vulnerable. As anger typically reflects vulnerability, competencies dissipate anger that may ultimately create conflict.

Douven et al. (2012) suggest that perhaps the most important work on professional competency is that of Cheetham and Chivers (1996), which focuses on four key components:

1　Knowledge/cognitive competence: the possession of appropriate work-related knowledge and the ability to put it into effective use, for example, theoretical/technical knowledge of hydrology and hydraulics, tacit knowledge, procedural knowledge of finances or projects and contextual knowledge of geography or technology

2　Functional competence: the ability to perform a range of work-based tasks effectively to produce specific outcomes, for example, occupation-specific skills such as report writing, IT literacy, budgeting and project management

3　Personal or behavioral competence: the ability to adopt appropriate behaviors in work-related situations, for example, self-confidence, control of emotions, listening, objectivity, collegiality, sensitivity to peers and conformity to professional norms

4　Values/ethical competence: the possession of appropriate professional values and the ability to make sound judgments, for example, adherence to laws, social/moral sensitivity and confidentiality

But different functions require combinations of competencies. Water professionals must have a mix of competencies that may even vary for the same occupation; these competencies may depend on a given cultural, socioeconomic and other professional setting (Douven et al., 2012; Uhlenbrook and de Jong, 2012). Individual competencies are sculpted by personalities, training and education received and professional experiences. As a consequence, teaching philosophies must now fit the new paradigm of the 'compassionate' water resources professional proffered by Berndtsson et al. (2005), who conclude that university curricula for water experts must establish strong links with the socioeconomic and human sciences. These include how to approach interest groups and decision makers, meet opposition and negotiate, act as educators and trainers and explain methods and visualize techniques in a 'pedagogical manner' in order to transfer knowledge.

Communication competency has many dimensions and communication itself has many modes of transmission. The objective of communication is to transfer a message from the sender to the receiver to create shared meanings. Daniels and Walker (2001) indicate that communication is a symbolic process where reality is fundamentally about meaning. Communication through synthesis focuses on building collective or nested knowledge to foster collective action (Brown, 2008).

Cultural competency skills are becoming increasingly important for water negotiations. Cultural competency goes beyond the shared learned behaviors and meanings as described by Daniels et al. (2012), and also refers to differences in negotiating styles, for example, between those from individualistic cultures versus those from collectivist or cooperative cultures. Negotiations research often uses nationality as a proxy for culture. But Daniels et al. (2012) indicate that a much more nuanced definition of cultures yields multiple cultures within a nation. The urban-rural divide is a good example of multiple cultures within a nation or region.

Online competency is a new area for water negotiations, and it goes beyond the traditional notion of functional skills needed for information technology and communication (ITC) and decision support systems to an online component commonly employed in cooperative or mediated modeling or morphological analysis. These fields are beyond the scope of this book; however, the reader will find good summaries of these topics in van den Belt (2004), Tidwell and van den Brink (2008), Giupponi et al. (2011) and Ritchey (2013). I experienced the importance of both cultural and online competencies first-hand over a period of three years as an online mediator for a contractor to eBay, the online auction company. Traditional conflict resolution training tacitly assumes that all mediation takes place between parties that can 'speak' verbally and interpersonally. Hammond (2003) indicates that online dispute resolution is effective because disputants behave online in much the same way as they do off-line.

Online competency has to be developed because the Internet is the fastest growing form of communication and is available worldwide (Wahab and Rule, 2003). As e-mail expands, it is also used to make unpopular requests and to avoid voicing complaints in person; for that reason, it is sometimes referred to as the 'coward's choice' (Ford, 2003). While e-mail enables unpleasant business to be dealt with at a distance, however, it also provides an opportunity to negotiate disputes between hearing-impaired persons or across distances or other gaps that may separate disputants (Hammond, 2003). As with face-to-face communication, the Internet presents new opportunities for confusion, misunderstandings, avoidance and, because there are no nonverbal cues, the potential for miscommunication increases (Ford, 2003).

According to C. W. Moore (2003), electronic dispute resolution has been adapted for a variety of disputes, including e-commerce, intraorganizational differences, insurance claims, family law and commercial contracts. Within the emerging field of electronic dispute resolution may be found a virtual ombuds office, real-estate contracts and environmental dispute resolution (Freid and Wesseloh, 2002). The role and function of the mediator varies from being primarily a technical manager of exchanging information to being highly influential in managing the negotiation process (C. W. Moore, 2003).

The importance of online competency for water conflicts is that many countries are just beginning the organization of alternative dispute resolution systems. Wahab and Rule (2003) have demonstrated that if nations incorporate online options early in the conflict resolution process, it becomes much easier to integrate these options into the national dispute resolution framework, in contrast to nations with 'conventional' dispute resolution programs that later attempt to integrate online options into their networks or 'toolboxes'. Emerging markets for online dispute resolution include India, Africa, Eastern Europe and Southeast Asia, as computers and operating systems become more available in developing countries. Over 60 providers of online dispute resolution services are listed on ODR.INFO, the home of the National Center for Technology and Dispute Resolution (NCTDR), and the primary portal for the field of online dispute resolution

(ODR). Looking forward on the role of ODR competency for negotiations in the legal industry, Susskind (2013, p102) writes,

> I predict that ODR will prove to be a disruptive technology that fundamentally challenges the work of traditional litigators (and of judges). In the long run, I expect it to become the dominant way to resolve all but the most complex and high-value disputes. For law firms and court lawyers, this is a direct assault on their conventional work. But it is also a great opportunity to become a leading player in this new, currently uncontested market space.

And while there remain challenges in online communication, Ford (2003) sees the likelihood that online dispute resolution 'will in time replace face-to-face conflict resolution and where it does not, will be used to complement the face-to-face process'. In *Water Diplomacy*, Islam and Susskind (2013) include this excerpted passage from 'Water Knowledge Networking: Partnering for Better Results' from J. Luiijendijk and W. L. Arriëns (2008, pp259–260), which echoes the importance of online competency: '[T]he prevailing digital divide in the world continues to impede access to such networking by many local practitioners and poor communities in the developing world, and also by the elderly who feel unable to participate. Unless water knowledge networks can find ways to bridge these divides, the risk is that the social capital of local traditional and indigenous water knowledge will be marginalized to extinction'.

Program targeted skills

Consider that the regional implementation organization for salmon recovery on the Columbia River is composed of at least nine federal agencies, several Native American tribes and five states. Next, consider how many entities are involved with reconsidering the storage, flood control and hydropower provisions connected to the Columbia River Treaty between the United States and Canada, with discussions regarding continuing or terminating the treaty beyond 2024. Or consider the groundwater situation in the Klamath Basin, located in southwestern Oregon and northern California, which consists of two state water departments – one of which manages groundwater and one that does not – three counties, four city municipalities, four active federal agencies, two agricultural organizations, 18 irrigations districts and approximately 4,000 residents. How is it possible to level the playing field for all of the programs associated with dams, fish, rivers, power, flooding, groundwater, ecosystems, law, environmental quality, navigation, recreation, irrigation and cultural resources, among many other issues, together for adaptive management?

Like the bubble level embedded in a compass case used to ensure a level playing field on Kai Lee's (1993) 'map' where science and politics mesh, collaborative learning as defined by Daniels and Walker (2001) draws upon systems thinking, public policy negotiation, conflict management and alternative dispute resolution to level the playing field amongst many competing interests and programs.

Collaborative learning approaches are well suited for natural resource, environmental and community decision-making situations that include (1) complexity, (2) controversy, (3) multiple parties, (4) multiple issues, (5) scientific and technical uncertainty and (6) legal and jurisdictional constraints (Daniels and Walker, 2012). The advantages of collaborative learning approaches to conflict management for this project include the following:

- It is learning-based public participation.
- Stakeholders learn from one another.
- Agencies (departments within municipalities) interact as stakeholders.
- Technical/scientific and traditional/local knowledge are respected.
- Public participation activities are accessible and inclusive.

Systems thinking enables one to see patterns of interdependency and to see into the future (Senge, 2012). Systems thinking is used to test policies and actions. In collaborative learning applications, systems thinking tasks are participatory and emphasize active learning, thus increasing knowledge and functional, personal and ethics or values-based competencies.

Daniels and Walker (2001, p186) explain the situation mapping tool as a key part of systems thinking: 'Situation mapping is the process of graphically representing a situation to create a shared and systemic understanding of it. The graphic, or spatial, depiction of the situation allows a far more relational understanding than could be developed through other means. It is quite a flexible technique; perhaps the only "rules" are as follows: (1) to put verbs on lines to convey the dynamic relationships and put nouns at the nodes of the lines to convey the elements in the system; and (2) to start in the middle of the page'.

In a retrospective following 20 years of application, Daniels and Walker (2012) summarize the lessons learned from the application of systems thinking to natural resources situations. Systems thinking (1) promotes a holistic understanding that is both accessible and pluralistic, (2) transforms a single issue focus into a multi-issue view, (3) clearly illustrates that complex situations cannot be fully managed/controlled, (4) corresponds well to natural resource management and (5) can encourage agencies to think beyond their default formulation of the situation paradigms that have emerged in the past 25 years. Finally, (6) systems thinking-based approaches require skilled facilitation.

We will investigate the use of situation maps and systems thinking in later chapters. The connection of collaborative learning and situation mapping to a holistic understanding of conflict situations is fundamental to the implementation of the Hydro-Trifecta Framework.

The graduated circle of transdisciplinarity

Compasses enable users to take bearings that can help them decide the direction to get from one point to another. Hydrogeologists use compasses to measure trends of geologic features such as folds, faults and fractures. Compasses have a graduated circle of degrees that permits accurate measurement of the bearing to navigate from a starting point to a destination. The graduated circle on the compass is divided into quadrants or an azimuthal system divided into 360 degrees.

The Hydro-Trifecta Framework relies on the philosophy of transdisciplinarity as a means to get a bearing on water situations because it is a more systemic and holistic way of seeing the world. Transdisciplinarity can be thought of as coordinating the various hierarchical levels that are traditionally used to address wicked problems, but are rarely linked together (Max-Neef, 2005). Lawrence (2010) completed a literature review on transdisciplinarity and summarized its shared aims, which include (1) offering a way of tackling complexity in science and knowledge fragmentation, (2) accepting local contexts and uncertainty by serving as a context-specific negotiation of knowledge, (3) implicating intercommunicative action and continuous collaboration during all phases of a project or 'mediation space and time', (4) being action oriented by making linkages across disciplinary boundaries, but also between the theoretical and professional practice – the bread and butter of the pracademic – and (5) bridging the gap between knowledge derived from research and decision-making processes in society.

Integrating a transdisciplinary approach to water negotiations is also the new paradigm for working across existing disciplines toward a new, higher-level synthesis – that is, to 'invent new science'. Because transdisciplinary research is grounded in 'real' problem situations, thereby involving stakeholder collaboration, it involves more fluid and evolving methodologies than are common in traditional academic research and it fosters mutual or transformative learning. The value of transdisciplinarity continues to be acknowledged as key to water governance. There remains limited guidance available on achieving it in practice, yet the next generation of water professionals has recognized it as a key development because it is more inclusive of other relevant fields. It necessitates additional competencies at a personal level, including the ability to engage with and value different cognitive and epistemological perspectives; the personalities and attitudes of stakeholders are as important as their discipline base and specialization (Patterson et al., 2013).

Consider, for example, the empirical level comprising the fields of mathematics, chemistry, physics, geology, ecology and sociology among others organized by the language of logics that addresses the questions of 'what exists'. The pragmatic level includes the technological fields of engineering, architecture, agriculture and medicine, organized by the language emphasizing the mechanical properties of nature and society that addresses the question of 'what we are capable of doing'. The normative level is composed of the planning, politics, law and environmental design built around the organizing language of planning that addresses the question of 'what we want to do'. And the value level goes beyond the present and the immediate, taking a bearing at the generations yet to come, at the planet as a whole, at an economy and is occupied by the fields of ethics, philosophy and theology; it addresses the question of 'what we must do' using the organizing language of deep ecology defined by a world that does not exist as a resource to be freely exploited by humans (Devall and Sessions, 1985; Max-Neef, 2005).

A parallel interpretation of 'what we must do' focuses on the world of decision making inherent in the social psychology of common pool resources, where a new superordinate identity is created, thereby expounding that 'we are all in this together' (van Vugt, 2009).

Sighting mechanisms using the Hydro-Trifecta compass needle

Compasses have a variety of sighting mechanisms to aid in taking bearings for deciding the direction of travel. The compass needle is magnetic and, in the absence of magnetic material near the compass, will point north. It is similar to Lee's (1993) notion of gyroscope to maintain an orientation through 'cross currents' of conflict. But is north always the best direction to travel or to use for a bearing? This is where the Hydro-Trifecta Framework provides the opportunity to set a bearing on a single path or to test multiple paths to see the world. Let's revisit the previously discussed water conflict transformation and negotiation frameworks listed here:

> *Water conflict transformation* focuses on *identity*, differentiating between rights, needs, benefits and equity using the Four Worlds/4i approach in the quest to create a new superordinate identity.
>
> *Water diplomacy* focuses on *interests*, on the flexible uses of water, and on joint fact finding to create value rather than zero-sum thinking through loops of societal, political and natural networks.
>
> *Water security* focuses on *investment and risk* and utilizes a web of climate, energy, food, water and community to define what might be tolerable for water use and reuse without getting into 'trouble'.

Acknowledging the importance of the 4Rs Framework developed by Dore et al. (2010), their framework overlaps and bridges the Water Security and Water Conflict Transformation frameworks. Paisley's (2008) important summary on an international water law framework appears to fit within well within the norms and values of the Water Diplomacy Framework.

Key features of the Hydro-Trifecta Framework

Throughout the development, refining and testing of the Hydro-Trifecta Framework, one of its key features has been that it is both integrative among the primary water negotiation frameworks and also customizable. The initial assessment of a water conflict situation during 'intake' to determine the potential for successful resolution is often considered part of the critical early phase of any intervention. With the Hydro-Trifecta Framework, however, stakeholder assessments to identify critical questions can be conducted at any time. This flexibility is important because conflicts often are not static, but can evolve and become more complex and chaotic through time, particularly as the conflict moves from the technical arena to the political arena. The most important interface in the

Hydro-Trifecta Framework involves its recognition that a water conflict may begin over efforts at joint fact finding associated with water diplomacy, then evolve to process and outcome equity or identity, factors typically associated with water conflict transformation. It then may get stuck on the issue of water security, as the definition of 'trouble' (for example, the investments and risks associated with water use and reuse) is defined and redefined under different climate-change scenarios that are related to parts of the water security web.

An application example – Nevada versus Utah groundwater development

The Hydro-Trifecta Framework provides a lens for examining the negotiations and renegotiations for the development of a shared aquifer system. Within the western United States, the carbonate rocks composing the Great Basin Aquifer serve as an excellent example of a transboundary aquifer that is in dispute. The Great Basin Aquifer underlying the states of Utah, Nevada, Idaho and Oregon is targeted for development by the Southern Nevada Water Authority (SNWA). This project focuses on the negotiation discourse from a variety of perspectives and factors. Data for this case study comes from a range of sources, including observations made while working in the area proposed for well drilling and pumping, conversations with scientists and lawyers from the state of Utah and published media reports. While the negotiations involve multiple discourses that compete for dominance, this example will focus on the technical and legal opportunities for common ground and agreement.

The city of Las Vegas depends on the Colorado River, specifically the Lake Mead reservoir, for about 90% of its drinking water. The SNWA is the water purveyor for Las Vegas and is currently constructing a third intake, costing $800 million, that will be 4.8 km in length and 183 m beneath the lake surface. The project was begun while Las Vegas was experiencing a decade of drought, and reservoir levels threatened to drop below the existing two shallower intakes. The SNWA is building flexibility into their water scarcity portfolio by planning to pump 140 million m^3 of groundwater, from over 140 wells located in four northeastern Nevada valleys. The water will be transported via a pipeline 423km in length to Las Vegas, and the anticipated cost is $15.5 billion (Smart, 2012). The project started in 1989 after research determined that groundwater is stored in a regionally extensive, but complexly deformed fractured-rock aquifer system. The Great Basin Aquifer is considered by some hydrogeologists as a regional aquifer system or 'megawatershed' that underlies the states of Utah, Nevada, Idaho and Oregon. The aquifer is targeted for development by the SNWA due to the apparent lack of hydraulic connection between groundwater flows in the nearly 260 surface watersheds that overlie the Great Basin Aquifer System.

Opposition to the project regularly filled the editorial pages of regional newspapers in both Nevada and Utah. The classical urban-rural divide was a discourse for both states, with ranchers and Native Americans in northern Nevada and western Utah arguing that pumping groundwater would impact the livelihoods and way of life for generations to come and would only benefit the urban landscape and casinos in Las Vegas. The potential impacts to troglodytes and stygofauna living within the caves and other karst features within the Great Basin National Park located near the proposed wellfields also featured prominently in the discourse (Brean, 2009). Public health concerns over dust blowing easterly to Salt Lake City from the dewatered basins were used to build coalitions against the project.

And as is common in the field of hydrogeology, working hypotheses of ground-water circulation are many in the region, and long-term safe yield and sustainable beneficial use are offered by consulting hydrologists, government scientists from the U.S. Geological Survey, the state of Nevada Desert Research Institute, the Utah Geological Survey and university professors from regional universities alike. In what is becoming a more frequent occurrence, a professor from a private regional university provides a discussion of their peer-reviewed technical journal article in a regional newspaper:

> The traditional view, the so-called interbasin flow model, maintains that the fractured limestones that make up much of the bedrock beneath the desert mountains and valleys of western Utah and eastern Nevada allow water to flow quite freely. In other words, water can flow beneath multiple valleys, and the mountains that separate them, as long as there is limestone to flow through. The interbasin flow model views limestone as one enormous aquifer. This view has two implications for Utah. First, water beneath Spring Valley, Nevada, should flow into Snake Valley, Utah; thus, water development in Nevada can impact water availability in Utah. Second, water development in such a large aquifer is like sucking water through little straws from a very large cup. You can 'drink' for a long time. Although the technical details are beyond the scope of an op-ed piece, our research strongly suggests that the interbasin flow model is largely wrong. Each basin behaves more-or-less independently. Water develop-ment is more like sucking water through straws from many small cups. . . . The impacts of water development cannot be anticipated until the way in which water moves, how much of it there is and how long it has been in the ground are understood. Until that day, Gov. Gary Herbert should not sign anything.
>
> (Nelson, 2012)

In negotiations over groundwater, a common practice is to have the experts prepare a jointly signed explanation. In this case, while Bredehoeft and Durbin (2009) acted as consultants to opposing sides of the groundwater development project, both parties prepared a peer-reviewed technical paper describing the groundwater modeling predictions.

The initial controversy funneled down to whether the federal Bureau of Land Management would permit the wellfield and pipelines on federal lands. But once these permits were granted in December 2012, the matter fell to the issue of water rights and water sharing. A congressional act, Public Law 108–424, authorized pipeline rights-of-way for the project, provided that Utah and Nevada divide the Snake Valley groundwater. Legal experts in western water law counseled the governor of Utah to sign a water-sharing agreement with the state of Nevada in

2013, which stated that the water in the valley was to be split in half between the states. Utah governor Gary Herbert rejected the agreement in April 2013 because 'Millard County Commissioners, White Pine County, the Great Basin Water Network, more than 350 additional plaintiffs in Nevada and Utah including environmentalists, ranchers, as well as the Church of Jesus Christ of Latter-day Saints (Mormon) and Confederated Tribes of the Goshute Nation used their lawyers to make both scientific and spiritual arguments opposed the agreement, fearing a replay of the devastation Los Angeles visited upon California's Owens Valley in the 1920s' (Maffly, 2013).

In a series of articles in the regional newspaper the *Salt Lake Tribune*, lawyers representing the Mormon Church at its large cattle ranch in nearby Spring Valley, used 'bequest value' arguments: 'A declaration of good intentions is all it is. There is no proof that it can do anything to avert disaster. . . . That's basically taking the future of our children and squandering it for short-term gain' (Maffly, 2013a). The lawyers for the Confederated Tribes of Goshute Reservation used both 'sustainability' and 'spiritual values' in their arguments when referring to the residents of the high desert range in eastern Nevada and western Utah, who 'understand the concepts of limits. . . . The concept that this water is lost and subject to capture completely ignores that people rely on those plants and springs' (Maffly, 2013a). The Shoshone and Goshutes also contended that the planned exportation of groundwater would desecrate a region that holds spiritual significance because the area is a source of traditional foods and medicines. The area also harbors the spirits of those slain in two U.S. Army massacres (Maffly, 2013a, 2013b).

Yet, the Utah Water Development Commission voted in May 2013 to encourage the governor of Utah to reopen talks with the state of Nevada. The reasoning for this appeared to focus on how the governor's decision regarding the shared agreement might affect the state of Utah's campaign to secure water rights, regulatory approvals and rights of way to build the Lake Powell Pipeline, a $1.5 billion water-supply project composed of a 220-km-long pipeline designed to divert approximately 106 million m^3 from the Colorado River, a river shared between Utah, Nevada and five other states (Maffly, 2013a, 2013b).

Applying the Hydro-Trifecta Framework

As previously outlined, the Hydro-Trifecta Framework is built on the foundation of a compass case of skills. These skills support the sighting mechanism for the modalities of negotiations, which point to the graduated circle of transdisciplinarity.

Scale targeted skills

It is clear that the scale of the Snake Valley project extends beyond the boundaries of the deep carbonate aquifers underlying the Great Basin of western Utah and eastern Nevada to the Colorado River Basin, given the apparent relationship between the groundwater pumping project desired by the SNWA and the proposed state of Utah Lake Powell Pipeline project.

Competency targeted skills

Both Nevada and Utah employed scientists and legal scholars with high levels of knowledge, functioning and professional competencies to develop the information for a draft agreement between both states. The Bureau of Land Management also prepared the environmental impact statement where 10 federal agencies, 14 state agencies and 8 local agencies were reportedly involved.

For example, the 2009 draft agreement between the U.S. states of Utah and Nevada over the allocation of the groundwater stored in the transboundary Carbonate Aquifer blurs the boundaries between both states: 'The States agree to work cooperatively to (a) resolve present or future controversies over the Snake Valley Groundwater Basin; (b) assure the quantity and quality of the Available Groundwater Supply, (c) minimize the injury to Existing Permitted Uses; (d) minimize environmental impacts and prevent the need for listing additional species under the Endangered Species Act, (e) maximize the water available for Beneficial Use in each State, and (f) manage the hydrologic basin as a whole' (Utah Department of Natural Resources, 2009).

However, the governor of Utah made his decision to reject the draft agreement based on values and ethical competencies. Rural leaders in western Utah; Millard County commissioners; environmental groups, such as the Center for Biological Diversity, Protect Snake Valley and the Great Basin Water Network, among over 70 other groups; local water users; and the Mormon Church all feared a repeat of a dust bowl, as well as lost opportunities for bequests and spirituality.

Clearly there was concern over some of the competencies of the teams responsible for the Draft Environmental Impact Statement (EIS), as it received over 400 comments covering a broad spectrum of topics that ranged from the technical to human resources. Some focused on the potential impacts to human health and well-being, including concerns related to subsidence associated with groundwater pumping, to the impacts of pumping-induced desertification and associated dust on human health and visibility. Related comments also focused on the potential impacts to the viewshed, recreation, tourism and visitation to the Great Basin National Park.

Although the Draft EIS listed nearly 20 tribal governments and even more tribal-related organizations that were involved with consultations, comments indicated there was inadequate tribal consultation. Native American concerns also related to loss of historic lands, tribal consultation policies (TCPs), artifacts, plants and animals of cultural importance and the loss of water that many tribes hold sacred.

There was also a perception that the project could foreclose future economic development opportunities in White Pine County, Nevada, and the Utah portion of the Snake Valley (Millard County).

Program targeted skills

Despite the required efforts of consultations, public meetings and opportunities for public comment, it was clear that the governor's office of the state of Utah, the Mormon Church, Native American tribes and environmental groups desired more information and learning beyond the groundwater, civil engineering, law and other topics addressed in the Draft EIS. The program development is clearly the classical 'Tech-Reg' approach, where agencies apparently employed the Three-I model – invite, inform and ignore – as described by Daniels and Walker (2001). Perhaps collaborative learning and situation mapping clearly depicting various linkages, and thus acknowledging these apparently silent voices, would have benefited the process of decision making in this regard. Some of the collaborative learning could have also benefited from the use of online video to share messages between entities who either could not attend meetings and consultations due to schedule or travel, or due to fear of intimidation.

Which siting mechanism should be used?

Water security?

Water security utilizes a web of climate, energy, food, water and community to define what might be tolerable risk for water use and reuse without getting into 'trouble'. Trouble for SNWA was defined in 1989 as needing adequate quantities of water for sustaining livelihoods, human well-being and socioeconomic development for the urban residents of Las Vegas. Water security for Utahns focused on assurances of socioeconomic development for rural residents living in the vicinity of Snake Valley, as well as protecting against the water-related disaster of a 'dust bowl' descending on the urban dwellers of Salt Lake City. The preservation of the viewshed and ecosystems associated with the Great Basin National Park was also considered part of their water security. As a consequence, there are some outstanding issues that need to be addressed for both states. Trouble for Utah also focuses on the need to secure water rights, regulatory approvals and rights of way to build the Lake Powell Pipeline for sustaining the livelihoods, human well-being and socioeconomic development for the residents of the St. George area.

Water diplomacy?

Recall that the Water Diplomacy Framework set its sights on the flexible uses of water and joint fact finding to create value, rather than zero-sum thinking through a loop of societal, political and natural networks. The value of continuing down this path in the Snake Valley situation is probably limited, as the states of Utah and Nevada not only developed a draft agreement, but there was also a congressional act designed to create value for all parties.

Water conflict transformation?

Recall that the water conflict transformation 'needle' is used to point disputants between the issues of rights, needs, benefits and equity, while at the same time moving beyond institutions, such as each state, toward securing information and creating incentives in the quest to create a new superordinate identity. This important approach was overlooked in the negotiations over the Snake Valley and there is value in going back to the negotiating table to redefine this core motive for decision making.

The graduated circle of transdisciplinarity

What direction is the best to go for the Snake Valley groundwater situation, given that the Utah governor rejected the shared water agreement due to concerns of rural residents, the Mormon Church, environmental groups and Native Americans over the spirituality of water? Recall that transdisciplinary partnerships move beyond existing disciplines to new higher level synthesis – that is, to support citizen or 'traditional' science and form alliances with other knowledge cultures.

What exists?

Recall that both states have much invested on the logic of the project including the geology, hydrology, ecology and numerical modeling of the groundwater system to learn 'what exists'. Opposing experts have jointly prepared reports. But there are still 'new' interpretations of how groundwater circulates through the Great Basin Aquifer, as discussed in the opinion pages of regional newspapers and articles in peer-reviewed journals. Guerrilla well-fare and hydrostitution as described by Jarvis (2010) will continue unabated. The age of the expert is over for this project.

What are we capable of doing?

The technological aspects of the project in terms of engineering clearly answer the question of 'what we are capable of doing'. There are two competing projects in the planning stages: the SNWA groundwater pumping and pipeline scheme and the Lake Powell Pipeline and Colorado River diversion scheme for Utah. Both regions have active citizen water conservation and water reuse programs to extend the flexibility of their respective water supplies.

What do we want to do?

The comprehensive environmental impact statement for the SWNA project and the shared water agreement rejected by the governor of Utah address the question of 'what we want to do'. The new wrinkle in the calculus of the SNWA project is the Lake Powell Pipeline project designed for southern Utah.

What must we do?

The question of 'what must we do' in terms of ethics, philosophy and theology remains problematic. And this is where the water security and water conflict transformation needles, using the Hydro-Trifecta Framework, should be used to guide a potential solution to the dueling projects.

Clearly, all parties associated with the SNWA and Lake Powell Pipeline projects live in a desert environment. While the developed groundwater would benefit the urban residents of Las Vegas, at the potential cost of threatening the public health of the urban residents in Salt Lake City, both states are located in the western United States, where water law dictates that water has value only if it is beneficially used. If Las Vegas does not develop the groundwater, how long will it be until Salt Lake City explores the development of the same groundwater? Likewise, if water levels in Lake Mead continue to drop due to regional drought and eventually compromise the SNWA intakes beneath the lake surface, is it possible that the 'Law of the River' – the compact negotiated between the seven states that share the Colorado River – will need to be renegotiated, thus risking the water levels in Lake Powell to the point the compromising the diversion associated with the Lake Powell Pipeline?

The common risk shared between both states, as well as the affected Native American tribes, is that of compromising the hydrologic integrity of Great Basin National Park and the Lake Mead National Recreation Area. Perhaps water security and the resolution of the perceived conflicts are best addressed by creating a new, superordinate identity – one that recognizes that, as residents of the desert, 'we are all in this together', and that Great Basin National Park and Lake Mead National Recreation Areas are worth preserving for all Nevadans and Utahns.

References

Abukhater, A. (2013) *Water as a catalyst for peace – transboundary water management and conflict resolution*, Routledge, Oxon, UK

Allan, J. A. and Mirumachi, N. (2010) 'Why negotiate? Asymmetric endowments, asymmetric power and the invisible nexus of water, trade and power that brings apparent water security', in A. Earle, A. Jägerskog and J. Öjendal (eds) *Transboundary Water Management: From Principles to Practice*, Earthscan, London, UK, pp13–26

Argyris, C. and Schön, D. (1996) *Organizational learning II: Theory, method and practice*, Addison Wesley, Reading, MA

Berndtsson, R., Falkenmark, M., Lindh, G., Bahri, A. and Jinno, K. (2005) 'Educating the compassionate water engineer – a remedy to avoid future water management failures', *Hydrological Sciences*, vol 50, no 1, pp7–16

Brean, H. (2009) 'New cave species have been identified at Great Basin National Park', *Las Vegas Review-Journal*, 4 October, www.reviewjournal.com/news/new-cave-species-have-been-identified-great-basin-national-park

Bredehoeft, J. and Durbin, T. (2009) 'Ground water development – the time to full capture problem', *Groundwater*, vol 47, pp506–514

Brown, V. A. (2008) *Leonardo's vision: A guide to collective thinking and action*, Sense Publishers, Rotterdam, the Netherlands

Cascão, A. E. and Zeitoun, M. (2010) 'Power, hegemony and critical hydropolitics', in A. Earle, A. Jägerskog and J. Öjendal (eds) *Transboundary Water Management: From Principles to Practice*, Earthscan, London, UK, pp27–42

Cheetham, G. and Chivers, G (1996) 'Towards a holistic model of professional competence', *Journal of European Industrial Training*, vol 20, pp20–30

Daniels, S. E. and Walker, G. B. (2001) *Working through environmental conflict: The collaborative learning approach*, Prager, Westport, CT

Daniels, S. E. and Walker, G. B. (2012) 'Lessons from the trenches: Twenty years of using systems thinking in natural resource, conflict situations', *Systems Research and Behavioral Science*, vol 29, pp104–115, doi:10.1002/sres.2100

Daniels, S. E., Walker, G. B. and Emborg, J. (2012) 'The unifying negotiation framework: A model of policy discourse', *Conflict Resolution Quarterly*, vol 30, no 1, pp3–31

Delli Priscoli, J. and Wolf, A. (2009) *Managing and transforming water conflicts*, Cambridge Press, New York, NY

Devall, W. and Sessions, G. (1985) *Deep ecology: Living as if nature mattered*, Gibbs M. Smith, Salt Lake City, UT

Dore, J., Robinson, J. and Smith, M. (eds) (2010) *Negotiate – reaching agreements over water*, IUCN, Gland, Switzerland

Douven, W., Mul, M. L., Álvarez, B. F., Son, L. H., Bakker, N., Radosevich, G. and van der Zaag, P. (2012) 'Enhancing capacities of riparian professionals to address and resolve trans-boundary issues in international river basins: Experiences from the Lower Mekong River Basin', *Hydrology and Earth System Sciences*, vol 16, pp3183–3197

Dovers, S. (2010) 'Embedded scales: Interdisciplinary and institutional issues', in V. A. Brown, J. A. Harris and J. Y. Russell (eds) *Tackling Wicked Problems through the Transdisciplinary Imagination*, Earthscan, Washington, DC, pp182–192

Ford, J. (2003) 'Integrating the Internet into conflict management systems', Mediate.com, www.mediate.com/articles/ford10.cfm

Foster, S. and Ait-Kadi, M. (2012) 'Integrated Water Resources Management (IWRM): How does groundwater fit in?' *Hydrogeology Journal*, vol 20, pp415–418

Freid, T. L. and Wesseloh, I. (2002) 'Integrating information technology into environmental treaty making' in L. Susskind, W. Moomaw and K. Gallagher (eds) *Transboundary Environmental Negotiation – New Approaches to Global Cooperation*, Jossey-Bass, A Wiley Company, San Francisco, CA, pp205–229

Giupponi, C., Mysiak, J., Depietri, Y. and Tamaro, M. (2011) 'Decision support systems for water resources management: Current state and guidelines for tool development', in P. A. Vanrolleghem (ed) *Decision Support for Water Framework Directive Implementation*, IWA Publishing, London, UK, pp107–202

Hammond, A. G. (2003) 'How do you write "yes": A study on the effectiveness of online dispute resolution', *Conflict Resolution Quarterly*, vol 20, pp261–286

Innes, J. E. and Booher, D. E. (2010) *Planning with complexity: An introduction to collaborative rationality for public policy*, Routledge, New York, NY

International Water Law Project (2013) 'International Water Law Project: Addressing the future of water law and policy in the 21st century', www.internationalwaterlaw.org/, accessed 2 September 2013

Islam, S. and Susskind, L. E. (2013) *Water diplomacy: A negotiated approach to managing complex water networks*, RFF Press, Routledge, New York, NY

Jarvis, W. T. (2010) 'Water wars, war of the well, and guerilla well-fare: Ground water', *Groundwater*, vol 48, no 3, pp346–350, doi:10.1111/j.1745–6584.2010.00695.x

Jarvis, W. T. (2012) 'Book review of *Water Diplomacy: A Negotiated Approach to Managing Complex Water Networks*', *Groundwater*, vol 50, no 6, p825

Jarvis, T. and Wolf, A. (2010) 'Managing water negotiations: Theory and approaches to water resources conflict and cooperation', in A. Earle, A. Jägerskog and J. Öjendal (eds) *Transboundary Water Management: Principles and Practice*, Earthscan, London, UK, pp125–141

Kennedy, K., Simonovic, S., Tejada-Guibert, A., de França Doria, M. and Luis Martin, J. (2009) 'IWRM implementation in basins, sub-basins and aquifers: State of the art review', the United Nations World Water Assessment Programme, Side Publication Series, United Nations Educational, Scientific and Cultural Organization, Paris, http://unesdoc.unesco.org/images/0018/001817/181790e.pdf

Lankford, B., Bakker, K., Zeitoun, M. and Conway, D. (eds) (2014) *Water security: Principles, perspectives and practices*. Routledge, London, UK, and New York, NY.

Lasswell, H. D. (1936) *Politics: Who gets what, when, how*, McGraw-Hill, New York, NY

Lawrence, R. J. (2010) 'Beyond disciplinary confinement to imaginative transdisciplinarity', in V. A. Brown, J. A. Harris and J. Y. Russell (eds) *Tackling Wicked Problems through the Transdisciplinary Imagination*. Earthscan, Washington, DC, pp16–30

Lee, K. N. (1993) *Compass and gyroscope*, Island Press, Washington, DC

Luiijendijk, J. and Arriëns, W. L. (2008) 'Water knowledge networking: Partnering for better results', in G. J. Alaerts and N. L. Dickinson (eds) *Water for a Changing World – Developing Local Knowledge and Capacity: Proceedings of the International Symposium, Delft, The Netherlands, June 13–15, 2007*, Taylor and Francis, London, UK, doi:10.1201/9780203878057.ch14

Maffly, B. (2013a) 'Rejecting Nevada water deal hurts Utah, critics say Snake Valley: Officials wonder if state's stance could complicate proposed Lake Powell Pipeline', *The Salt Lake Tribune*, 25 May

Maffly, B. (2013b) 'Herbert not budging on Snake Valley deal, water managers say outlook dire without Lake Powell Pipeline', *The Salt Lake Tribune*, 18 June, www.sltrib.com/sltrib/news/56478014-78/utah-nevada-agreement-valley.html.csp

Matsumoto, K. (2009) 'Appendix E: Treaties with groundwater provisions', in J. Delli Priscoli and A. T. Wolf (eds) *Managing and Transforming Water Conflicts*, Cambridge University Press, New York, NY, pp266–273

Max-Neef, M. A. (2005) 'Foundations of transdisciplinarity', *Ecological Economics*, vol 53, pp5–16

McCaffrey, S. (2007) *The law of international watercourses*, Oxford University Press, Oxford

Moore, C. W. (2003) *The mediation process*, Jossey-Bass, A Wiley Company, San Francisco, CA.

Nelson, S. (2012) 'Nelson: Understand groundwater', Op-Ed, *The Salt Lake Tribune*, 17 November, http://archive.sltrib.com/article.php?id=24090887&itype=storyID

Oregon State University (2013) 'Transboundary freshwater dispute database', www.transboundarywaters.orst.edu/

Paisley, R. K. (2008) 'FAO training manual for international watercourses/river basins including law, negotiation, conflict resolution and simulation training exercises', Food and Agriculture Organization of the United Nations (FAO)

Patterson, J. J., Lukasiewicz, A., Wallis, P. J., Rubenstein, N., Coffey, B., Gachenga, E. and Lynch, A. J. J. (2013) 'Tapping fresh currents: Fostering early-career researchers in transdisciplinary water governance research', *Water Alternatives*, vol 6, no 2, pp293–312

Ritchey, T. (2013) 'Wicked problems: Modelling social messes with morphological analysis', *Acta Morphologica Generalis*, vol 2, no 1, pp1–8

Rittel, H. and Webber, M.M. (1973) 'Dilemmas in a general theory of planning', *Policy Sciences*, vol 4, pp155–169

Rothman, J. (1997) *Resolving identity-based conflict in nations, organizations and communities.* Jossey-Bass, A Wiley Company, San Francisco, CA

Senge, P. (2012) 'Systems citizenship: The leadership mandate for this millenium', *Reflections*, vol 7, no. 3, pp1–8

Shah, T. (2009) *Taming the anarchy: Groundwater governance in South Asia*, Resources for the Future Press, Washington, DC

Smart, C. (2012) 'BLM poised to OK Las Vegas plan to pump and pipe desert groundwater'. *The Salt Lake Tribune*, 3 August, www.sltrib.com/sltrib/politics/54624691-90/nevada-blm-final-eis.html.csp

Susskind, R. (2013) *Tomorrow's lawyers: An introduction to your future*, Oxford University Press, Oxford, UK

Tidwell, V.C. and van den Brink, C. (2008) 'Cooperative modeling: Linking science, communication, and ground water planning', *Groundwater*, vol 46, no 2, pp174–182

Tindall, J.A. and Campbell, A.A. (2012) *Water security: Conflicts, threats, policies*, DTP Publishing, Denver, CO

U.S. Department of Homeland Security (2013) 'Water and wastewater systems sector', www.dhs.gov/water-and-wastewater-systems-sector, last accessed 2 September 2013

Uhlenbrook, S. and de Jong, E. (2012) 'T-shaped competency profile for water professionals of the future', *Hydrology and Earth Systems Sciences*, vol 9, pp2935–2957, www.hydrol-earth-syst-sci-discuss.net/9/2935/2012/hessd-9-2935-2012-print.pdf

Utah Department of Natural Resources (2009). 'Agreement for management of the Snake Valley Groundwater System', http://naturalresources.utah.gov/pdf/snake_valley_agree.pdf

Van den Belt, M. (2004) *Mediated modeling: A systems dynamics approach to environmental consensus building*, Island Press, Washington, DC

Van Vugt, M. (2009) 'Triumph of the commons: Helping the world to share', *New Scientist*, no 2722, pp40–43

Wahab, M.A. and Rule, C. (2003) 'The leading edge: On-line dispute resolution in the developing world', *ACResolution*, Summer 2003, pp32–34

Wolf, A.T. (2008) 'Healing the enlightenment rift: Rationality, spirituality and shared waters', *Journal of International Affairs*, vol 61, no 2, pp51–73

Wolf, A.T. (2012) 'Spiritual understandings of conflict and transformation and their contribution to water dialogue', *Water Policy*, vol 14, pp73–88

Wouters, P. (2013) *Water security and the global water agenda: A UN-water analytical brief*, United Nations University, Ontario, Canada

Zeitoun, M. (2011) 'The global web of national water security', *Global Policy*, vol 2, no 3, pp286–296

4 The silent revolution and the coming war of the wells?

> An easily understood, workable falsehood is more useful than a complex, incomprehensible truth.
>
> – Thumb's second postulate (from Schwartz, 2013)

Whereas the classic disputes over resources are usually over the territorial integrity and the extent of government control, as described by Anderson (1999), Thomasson (2005) posits that it is incompatibilities over resources that create grievances or conflicts, and that 'scarce' does not have to mean that the resource is limited. Incompatibilities often arise over the use and equitable, or inequitable, distribution of a resource and the competing values associated with its use (Tidwell and van den Brink, 2008).

State, provincial and county agencies and citizen-based groups rely on assessments by groundwater experts for water well siting, land development and drinking-water supply protection programs. County agencies and citizen-based groups often retain additional experts to conduct 'peer reviews' of the reviews and studies prepared by hydrogeologic experts on behalf of their clients. In addition to the incompatibilities over use, equity in process and outcomes, conflicts over groundwater are unique in that the distribution of groundwater, 'rights' and 'values' attached to groundwater, conceptual models and uncertainty, as well as missing information, inaccurate data and how the 'science' will be interpreted and used by the different experts, fuel 'contested expertise' and 'dueling expert' syndromes. Compounding these incompatibilities are 'hydromyths' that exist among water experts and groundwater managers. The traditional hydromyths include that groundwater is an unreliable or fragile resource and that groundwater mining is always unethical because it is unsustainable and damages future generations.

This chapter discusses three issues that hydrogeologists are increasingly facing in their work: (1) 'contested expertise' from folk beliefs practiced by water experts from outside the sciences who site water wells, (2) expert opinions on the suitability of water supplies from low-capacity wells that are 'exempt' from permitting or licensing and (3) the 'dueling experts' situation that often emerges when hydrogeologic assessments are completed. Two case studies are used to show how contested expertise and dueling experts situations related to water wells evolve and that they are not easily resolved, if at all.

War of the well?

Most water professionals are familiar with transboundary disputes over surface water; the media and academic journals are replete with national and international examples. Although groundwater disputes are less well known, they are increasingly becoming newsworthy. These disputes involve more than quantity, quality and distribution and include participation from many scientific disciplines, special interest groups and the public, as well as the influence of folk beliefs. Until 2004, conflicts over transboundary groundwater generally focused on the contamination of wells (Gleick, 2009). Yet concerns over access to water in drought-prone regions such as Somalia have given rise to a new generation of conflict over groundwater. Gleick and Heberger (2012) chronicled a duel wherein two farmers in Mexico killed each other over the rights to a spring, as well as clashes over wells in Yemen and Kenya, including the destruction of a drilling rig. A 'war of the well' made it into the mainstream media when the *Washington Post* reported about two neighboring clans in Somalia in 2006. In that report, a Somali nurse with International Medical Corps, a nonprofit relief group, was quoted as saying that the dispute was 'like the start of the water wars right here in Somalia' (Wax, 2006).

In the United States, the stakes are a little higher than battles over individual wells and springs. The state of Mississippi filed a lawsuit against the City of Memphis, Tennessee in 2009 for capturing groundwater stored in the Memphis Aquifer underlying Mississippi and is seeking $1 billion in damages (Cameron, 2009). Likewise, the states of Utah and Nevada continue to negotiate over water stored in a shared, fractured-rock aquifer, which will serve as part of the municipal water supply for the City of Las Vegas, Nevada. As discussed in the previous chapter, the water will be conveyed through a 423-km-long pipeline with the total costs of the project approaching $15.5 billion. The viability of the project remains in play, however, as the Governor of the state of Utah rejected the agreement for sharing the groundwater with the state of Nevada.

The silent revolution

A 'silent revolution' is occurring where millions of farmers are pursuing short-term benefits associated with the intensive use of groundwater for agricultural use in India, China, Mexico and Spain and where proactive governmental action is needed to avert water conflicts between neighboring users, user groups, states, provinces and nations (Llamas and Martínez-Santos, 2005a, 2005b). A comparable situation exists with permit-exempt wells typically reserved for domestic, stock and garden use in many states within the western United States, Canadian provinces, Australia and some European countries. As listed in Table 4.1, the number of wells or shallow 'water extraction mechanisms' has increased exponentially in many parts of the world as dramatic changes in drilling technology, pumping technology and the availability of electrical and diesel power has increased over the past 60 years.

Table 4.1 Representative numbers of wells in select countries

Country	Number of wells
India	19–26 million 'wells' or water extraction mechanisms (WEM)
United States	15.9 million
China	3.4–3.5 million
Canada	1,000,000
Germany	500,000
South Africa	500,000
Iran	500,000
Pakistan	500,000
Spain	500,000
Mexico	96,000
Denmark	75,000
Taiwan	37,000
Mongolia	27,000
Ireland	10,000
Malta	10,000
Costa Rica	5,000

Exempt wells

Land-use planning commissions and staff must rely on the assessments of hydro-geologists to determine whether a tract of land subdivided into individual lots for rural housing, often referred to as a subdivision, can be served by groundwater resources underlying the proposed development. In his landmark address arguing for the passage of Senate Bill 100 in 1973, former Oregon governor Tom McCall referred to rural developments that are typically platted without hookups to municipal power, sewer and water sources and are located beyond power lines and paved roads as 'sagebrush subdivisions'.

Domestic wells in many western states and Canadian provinces are exempt from permitting requirements because of the perception that use from these wells is de minimis – that the quantity of water used by the random rural home is too small to measure or manage (Glennon, 2002; Bryner, 2004; Van de Wetering, 2007; Bracken, 2010) and that the transaction costs for the institutional management of these wells is considerable (Shah, 2009).

Domestic wells and some industrial- and commercial-use wells are 'exempt' from state controls, except for requiring the driller or landowner to 'notify' the jurisdictional authorities (Van de Wetering, 2007). All states in the western

United States, except California and Utah, as well as all western Canadian provinces exempt domestic wells from regulation (Bryner, 2004; Nowlan, 2005; Trout Unlimited, 2007; Bracken, 2010). And although the use from each household well may be small, the increase in the number of exempt wells in some regions has resulted in measurable depletions of groundwater (Trout Unlimited, 2007; Van de Wetering, 2007).

The other problems associated with exempt wells focus on (1) the fact that many land owners believe that the ability to install a well and withdraw water is a basic property right (Bracken, 2012) and (2) whether there should be some metric employed to determine if there are sufficient quantities of usable quality water for the proposed subdivisions and exurban area developments – because if water problems arise, land-use planners and commissioners are usually the first people disgruntled landowners visit with those problems. Hydrogeologic experts are recruited by developers, landowners and government agencies to provide input on locating and assessing the sustainability of groundwater sources. Each 'expert' has a different interpretation of the existing hydrogeologic information associated with the various facets of the land-use proposal, which can lead to the classic situation of contested expertise and 'dueling experts'.

Contested expertise

Hydrogeologists continually tread the sensitive – even explosive – interface between folk beliefs and scientific rigor as they attempt to solve problems in their dealings with others who don't conceive of, understand or relate to the world as hydrogeologists do. There may be conflict as water witches and dowsers contest the expertise of hydrogeologists and vice versa. There is tremendous steadfastness inherent in folklore that 150 years of quantitative hydrogeologic science has been unsuccessful in displacing. Schwartz (2013) indicates that getting beyond folk beliefs is difficult because there are often examples within the varied and complex earth systems sciences that fit these beliefs.

Dowsing and water witching is a global practice, despite considerable skepticism. Dowsers, water diviners and water witches are considered local groundwater experts. In 2008, the *UAE National Newspaper* describes an 'ab shanaas', the Afghan version of a water diviner, working outside Kandahar. Chevalking and others (2008) report that Pakistani dowsing experts are paid at least $500 to find groundwater. The August 3, 2007 edition of the *Wall Street Journal* ran a piece entitled 'In Race to Find Water, It's Science vs. "Witchers"' that detailed how a California 'dowser' charged $200 per hour, plus $10 for each gallon per minute produced from a well he had located, and that he sometimes made $7,500 in a day's work.

In the *National Driller's Journal*, editorials published in 1999 called water dowsing 'bad news' for groundwater, yet conceded, 'Obviously it is an issue we must be cognizant of to be effective in our business, and to be better communicators with dowsers and our clients' (Stone, 1999). Historical notes published in a 2002 issue of *Groundwater* softened the criticism on dowsing by indicating that,

While hydrogeologists have a better track record at finding water, we must also win the client's favor and trust', implying that people needing wells are more likely to believe in the success of a site located by a water witch than by a geologist. Birkenholtz (2008a, b) describes a comparable situation in north-western Rajasthan, India where farmers rely on the knowledge of traditional water diviners and have a deep-rooted mistrust of state engineers; the former are trusted not only because they are 'more accurate and cheaper', but also because 'their knowledge and technical knowhow is derived in the same way as that of the farmer, through practice in the area.

(Deming, 2002)

Chevalking and others (2008) provide a short history of dowsing in which they reference cave paintings dating from 9,000 years ago in the Atlas Mountains of North Africa and in Peru that depict men holding forked dowsing sticks; Confucius (2500 BC), who mentions dowsing in his writings; and the Egyptians, who show evidence of this practice in their stone drawings and carvings. Colleague Amitangshu Acharya shared with me the *Brihat Samhita* by Indian mathematician and astrologer Varahamihira (505–587 CE), which has a total of 4,000 slokas (hymns) with detailed information on the art of water divining in Chapter 53. The first sloka observes that there are sources of water, 'just as there are veins in the human body, even so do they exist, some higher up, others lower down, in the earth'.

Ellis (1917) indicates that the first documented use of dowsing for mineral exploration occurred in the late 1400s to locate minerals in Germany. According to *Le Fil de l'Eau*, a Pompidou Center publication and summarized by Bird (1993), the French were apparently the first Europeans to adopt dowsing in the search for water, together with perhaps the first mention of multi-tasking as we know it today by also using dowsing to search for 'murderers and Protestants'.

Dowsers use a variety of tools such as scissors, coconut shells, 'Spanish needles', pliers, welding rods, coat hangers, lead pipes and the well-known forked stick. Office supplies, such as rubber bands, and sophisticated remote sensing equipment are also used (Applegate, 2002). In the article 'Witching for Water in Oregon' published in a 1952 edition of *Western Folklore*, Claude Stephens describes the diverse skills and tools used by water witches dating back to the early 1900s. Stephens also reported an interesting array of equipment used in the hunt for Oregon groundwater, including one water locator's toolbox containing 'a compass, several copper rods, the forked witch stick, a radio tube, a small bottle of water, several stakes, a block of wood, and a gold watch and chain . . . (and) a rawhide covered buggy whip' (p204).

'Finding' groundwater is considered by many to be a gift endowed to those with powers of magical divination (Vogt and Hyman, 1979). In Europe, dowsing is considered a learned skill (Applegate, 2002).

Water witching and dowsing confront the professional hydrogeologist with the paradoxes of dealing where faith-based and science-based ideologies intersect in the real world. Conflicts can arise in the daily intersection of ideological divides. Such conflicts can be intractable if the respective camps find themselves pitted against each other, especially in front of a paying client.

California case study

I was requested to review several hydrogeologic reports prepared for a wealthy landowner whose ranch was located in Southern California. This area is famous not only for the many movie stars and pop singers living in the area, but also for the complex geology composing the California Coastal Range, with many folds and faults contorting the strata into vertical to overturned orientations. Over a couple of days, I selected potential drilling sites – and there were very few because of the complex geology and the clayey composition of the bedrock.

I returned to the ranch a couple of weeks later, at the request of the site manager, to look over a few drilling locations. My lack of enthusiasm for drilling large-capacity wells in vertical strata was quickly recognized by the site manager as I explained the geology and the uncertainty of production in such a complex setting. At every location where we stopped was a small garden of surveyor's pin flags. I asked about the purpose of the pin flags, only to learn that these marked the potential drilling locations selected by an out-of-state water witch.

On the basis of past experience with clients who relied on water witches to locate wells, I quickly deferred to the witched locations, but offered to show alternative locations in case the witched sites were unsuccessful. On the way to an alternative location, I asked to see the water witch's report and was handed a piece of paper about the size of a cash register receipt with listed depths where water would be encountered. The depth of the deepest reading was 329 m.

Drilling the chosen witched location proved to be very challenging as the greenish-gray, clay-rich serpentinite bedrock was tilted vertically. The well driller contacted me when the drill reached 300 m and was concerned about getting the drill pipe permanently stuck in the borehole due to the slow, sticky drilling conditions. After we discussed the pros and cons of continuing to drill, along with the costs associated with losing the drill pipe in the borehole, the borehole was abandoned – after nearly $100,000 had been spent. The landowner contacted the water witch about the lack of water at the 'witched' depths, only to be told that the driller had not drilled to the target depth. The landowner was not satisfied with the water witch, the well driller or the geologist and refused to pay any invoice until a reasonable explanation was provided for the dry hole situation.

Throughout the 1980s and including while I was attending college, Dr. Jay Lehr was the editor of the National Ground Water Association journals *Ground Water* and *Water Well Journal*. Dr. Lehr was not kind to water witches in his editorials. He offered 'Dowser Buster' T-shirts for sale in 1987. My hydrogeologic training was obviously influenced by these technical journals.

Over the years, I amassed quite a collection of books and literature on water witching and dowsing because I encountered many questions regarding the practice from clients and well drillers. The U.S. Geological Survey (USGS) surely has received many more inquiries over the decades, given that Water Supply Paper 416, published in 1917, was titled 'The Divining Rod: A History of Water Witching' (Ellis, 1917). The USGS did not provide a ringing endorsement of dowsing. Bird (1993) chronicles the 10-year debate between the USGS and the American Society of Dowsers over the objectivity of the USGS in their public information brochures.

Recalling the clayey rock that the driller encountered at the ranch prompted me to dig a little deeper in my library for an answer. In *The Complete Book of Dowsing*, Applegate (2002) reports that clay-rich rocks can interfere with the dowser's 'reading' of a landscape. I prepared a short report describing the drilling conditions and the clay-rich drill cuttings, along with a quote from Applegate (2002), and the landowner was satisfied with the explanation. All invoices were paid in full.

Applying the Hydro-Trifecta Framework

Although this conflict occurred just as I was finishing my academic training in conflict resolution, there is value in looking back at the situation through the lens of the Hydro-Trifecta Framework. The modalities of negotiations would work well in this situation and pointed to transdisciplinarity in the quest for synthesis leading to 'new' science – and the payment of the outstanding invoices.

Scale targeted skills

The drilling location situation touched upon both an interpersonal and inter-sectoral conflict between the geologist, water witch, driller and landowner. It would have been a simple matter to let the matter escalate to litigation, given the amount of money involved.

Competency targeted skills

On the basis of the existing information, it is clear that the various parties had a high level of knowledge, functional and values competencies. Personal or behavioral competence was one-sided, as the landowner and water witch were good listeners, but were not objective in their beliefs. It was important for both the geologist and the driller to recognize the beliefs of the landowner and water witch. Cultural competency and identity were important factors in resolving the conflict.

Program targeted skills

The common thread to this situation is that each 'water expert' came in differ-ent form – scientist, well drillers and water witch with training or experience in groundwater resources and the wealthy landowner – with each expert serving as a catalyst in the conflict and each seasoned a little differently by their relation-ship with each other. It was clear to me that the landowner and water witch were not going to change their beliefs regardless of how much 'science' I attempted to throw at them. It was up to the well driller and me to learn more about their folk beliefs.

Which siting mechanism should be used?

Water security?

This siting mechanism primarily addresses risk and utilizes a web of climate, energy, food, water and community to define what might be a tolerable risk for water use and reuse without getting into 'trouble'. While there was significant financial risk associated with drilling the borehole, water was not being used or reused until the well was drilled and completed.

Water diplomacy?

This siting mechanism acknowledges that water crosses multiple domains and multiparty negotiation and coalitional behavior. It primarily addresses interests and sets its sights on the flexible uses of water and joint fact finding to create value – also known as the 'mutual gains approach' – rather than the 'one party wins, one party loses' approach. Given that the dispute focused on getting unpaid invoices paid, the water diplomacy approach would have been well suited to this situation, as it attempts to synthesize explicit (scientific) and tacit (contextual) water knowledge. Joint fact finding and a search for nonzero sums options are important attributes of this negotiation framework. However, it is not clear where the conflict between empiricism and folk beliefs would fit within this modality.

Water conflict transformation?

This siting mechanism primarily addresses identity-based conflicts. It could be argued that the folk beliefs and identity associated with water witching were the principal cause for the conflict, and it is clear that the values of the geologist, the landowner and the water witch could have contributed to more distrust between the parties of this dispute. Their individual goals extended beyond advancing concrete interests to include upholding their dignity and their reputations. Once it was recognized that the invoices would not be paid unless the dignity and reputation of the parties were set aside, the interests of all parties were more focused on joint fact finding in order to appease the landowner.

The graduated circle of transdisciplinarity

Here we should recall that transdisciplinary partnerships move beyond existing disciplines to new, higher level synthesis – that is, to support citizen or 'traditional' science and form alliances with other knowledge cultures.

What exists?

The geologist, well driller, water witch and, to a greater extent, the land owner have invested much on interpreting and learning more about the geology and hydrology of the ranch in order to learn 'what exists'. Additional investment in garnering additional information at the dowsed site would not have contributed to new science

What are we capable of doing?

Given the geologic, hydrologic and drilling engineering conditions at the bore-hole site, the decision by all parties, save the water witch, was to stop drilling before incurring large expenditures associated with a stuck drilling pipe.

What do we want to do?

Clearly the options were limited to either stopping drilling and moving to a new location or moving toward litigation. But none of the parties desired to pursue litigation. The goal was to come to a consensus on an answer to the questions posed by the landowner: what was the reason for the differences in the conceptual model of the hydrogeology held by the water witch and what was encountered during drilling?

What must we do?

It was clear that the answer to the landowner's question needed to acknowledge water witching. While the traditional knowledge of the water witch could not be ignored, it was also unlikely that the water witch would change beliefs in order to accommodate a new answer to the existing problem. The geologist synthesized the 'new science' by integrating the results of the water witch's assessment with the known geology and hydrology based on the well driller's work.

Spaghetti western water war – Oregon case study

I live next to a Measure 37 claim which, if allowed, might very well 'take' my water. The area I live in is a designated 'Water Limited Area'. If my neighbor wins, I lose. Where is the justice in that? My neighbor, whom I've known for 20 years, is now my adversary. Measure 37 hasn't solved anything; it just pits neighbor against neighbor, lawyer against lawyer and makes the courts my land-use planning protection.

– Ted Gaty, resident of Salem, Oregon, letter to the editor, *Statesman Journal*, May 9, 2007

The saga over exempt wells in the western United States and Canada epitomizes a new type of water conflict – the 'spaghetti western water war', as described by Vinett and Jarvis (2012). In the 'spaghetti western', 'the herd' is in the millions and includes diverse 'breeds' of well water users: domestic, agricultural (farm and ranch) and industrial. In the classical sense of the spaghetti western film genre, the language of exempt wells is one that is difficult to translate from state to state, province to province and nation to nation. The political melodrama of the exempt well offers land developers and farmers a low-cost approach to providing water supplies to high-priced exurban or 'sagebrush subdivisions' found on the outskirts of cities, as well as to millions of small acreage farms. To compound the difficulty for water managers, most administrative bodies do not know the exact number of exempt wells in their respective jurisdictions, the locations of the wells or how much water they actually extract (Bracken, 2010). The highly fluid, emotionally charged storyline is cast with fading and rising stars in consulting and legal firms, well drillers, water diviners and 'hydrostitutes' marshaled by local governments and the courts, each dueling with the other, fueling the appetite of the 'hydrohydra' – the mythical multiheaded beast of the underground that feeds on conflicts over groundwater, as described in earlier chapters. Vinett and Jarvis (2012) describe how intractable the conflicts associated with this new genre of water war can be.

This case study used consultant reports, peer review reports, interviews with concerned citizens, relevant articles from the hydrogeologic literature and audience responses to the authors' presentations at professional and public forums. Newspaper articles and letters to the editor were also integrated into the analysis because (1) the United States and Canada are representative democracies where 'we can count on the media to help citizens educate themselves about the issues being deliberated by their elected representatives' (Susskind, 2005, p149) and because (2) the media can also 'fuel' a water dispute, as described by Putnam and Peterson (2003) for the Edwards Aquifer, by casting a conflict as between urban and rural factions. Likewise, Susskind (2005, p149) argues that the 'increasing technical complexity and often confusing interaction among decisions at multiple levels of government far exceed the capacity of most news outlets to explain what is going on and what various policy options imply'.

A case study in Oregon is ideal for analysis because the state land-use planning system was developed over 30 years ago and has served as a model for land-use planning to protect farms and forests across the globe (Knapp and Nelson, 1993). Oregon also pioneered an initiative and referendum process called the 'Oregon System' of direct legislation in the early 1900s that enabled citizens who gathered a sufficient number of signatures to place proposals on the ballot (Peterson del Mar, 2003). Ballot Measure 37 was introduced to Oregon voters in 2004; it proposed to permit landowners to apply for exemptions from land-use laws enacted after they purchased their land or to seek compensation for lost opportunities to develop their lands if the land-use rules remain in place. Ballot Measure 37 passed in 2004, upsetting the balance of the 19 planning goals that Oregon developed over the past 30 years for land use. The success of Ballot Measure 37 spurred property rights activists in many other states to propose comparable measures in order to weaken or cancel land-use laws. (For example, Ballot Initiative I-933 was a property rights initiative in Washington State comparable to Oregon's Measure 37 that was introduced in 2006. It lost by 59% of the vote. In the same year, the Private Property Rights Protection Act, introduced in Arizona under Proposition 207, passed by 64% of the vote.)

Over 7,500 applications for exemptions were submitted to local and state government agencies for review. According to the Measure 37 Database maintained by the Institute of Portland Metropolitan Studies at Portland State University,

72% of the claims were within the Willamette River Basin. Approximately 90% of these claims were under the jurisdiction of counties (Portland State University, 2013).

The vast majority of Measure 37 claims were located in exurban areas outside of urban growth boundaries. The claims, if approved and built as outlined in the applications, would have resulted in low-density, large-lot houses relying on individual 'exempt' wells rather than by community water systems found within the growth boundaries. According to Mortenson (2008), an estimated 126,000 new houses and potentially new exempt wells would have been built in rural areas under Measure 37. For comparison, the Oregon Water Resources Department (OWRD) reports approximately 250,000 exempt wells across Oregon with about 100,000 wells located in the Willamette River Basin; approximately 3,800 new wells are drilled annually (Vinett, 2011). Regulatory approval of all Measure 37 claims had the potential of increasing the number of exempt wells in the Willamette River Basin by 90% or by approximately 90,000 new wells.

Yet in November 2007, Ballot Measure 49 was introduced to scale back development associated with Measure 37. Measure 49 allowed claimants to build a few homes, but prohibited commercial and industrial development, as well as large subdivisions, and limited the number of exempt wells that could be installed by individual claimants in state-recognized groundwater-limited areas. Measure 49 was passed by a majority of Oregon voters. Mortensen (2008a, 2008b) reports that the estimated number of new houses and potentially new exempt wells that could be built in rural areas under Measure 49 approached 13,000.

But Pruesch (2008) reports that Oregon property rights activists are organizing 'to repeal Measure 49 and full implementation of Measure 37'. Roundtables convened by Oregon State University in 2008, which were organized to listen to the concerns of Oregonians as the state embarked on a statewide water planning effort, determined that exempt wells are a concern to water rights holders and land use planners (see Institute for Water and Watersheds, 2013).

Likewise, an international conference titled 'Exempt Wells: Problems and Approaches in the Northwest' was convened as a collaboration between several universities located in the western United States (Exempt Wells, 2011). On the basis of comments from the approximately 70 attendees from across the United States and Canada, issues with exempt wells are growing rapidly. The goal of the conference was to identify the critical issues associated with the management and impacts of exempt domestic wells, and to stimulate new ideas to solve the conflicts that have arisen between traditional water rights holders, well-drilling contractors and water users that rely on exempt domestic wells. It is clear that the discussion on exempt wells remains on the immediate horizon for groundwater planning in states and provinces across North America, as the energy economy ebbs and flows, and with rural subdivision development cycling along with the associated economic boom and bust cycles.

According to the OWRD, exempt uses of groundwater 'include single or group domestic use up to 15,000 gallons per day (gpd), non-commercial irrigation of up to one-half acre, stock watering and commercial and industrial use up to 5000 gpd' (OWRD and DLCD, 2002). Careful examination of Table 4.2 reveals that

Table 4.2 Summary of exempt well regulations in Western North America

Western state or province	Exempt use?	Conditions and comments
Alaska	Yes	• Any use less than 5,000 gpd (19 m³pd); or • Any use less than 500 gpd (2 m³pd) over a 10-day period; or • Non-consumptive uses less than 30,000 gpd (113 m³pd).
Arizona	Yes	• Any use less than 35 gallons per minute (gpm) (0.13 m³pm).
California	No	• No permit required. • No statewide groundwater policy. Groundwater typically managed by county.
Colorado	Yes	• Domestic use less than 15 gpm (0.06 m³pm). • Irrigation less than 1 acre and less than three families. • Stockwatering. • Commercial drinking and sanitation water use less than 15 gpm (0.06 m³pm). • Fire protection. • Observation and monitoring.
Idaho	Yes	• Domestic use exempt less than 13,000 gpd (49 m³pd). • Any use less than 0.04 cubic feet per second (cfs; 7.48 gpm; 0.4 m³pm) and less than 2,500 gpd (9.5 m³pd). • Irrigation less than 0.5 acre. • Livestock watering. • Organization camps and public campgrounds.
Montana	Yes	• Any use less than 35 gpd (0.13 m³pd) and less than 10 acre-ft per year (8,927 gpd; 34 m³pd). • Livestock less than 30 acre-ft per year (26,782 gpd; 101 m³pd).
Nevada	Yes	• Domestic less than 2 acre-ft per year (1785 gpd; 7 m³pd).
New Mexico	Yes	• Permit required, 'excepted' from denial of permit by state engineer. • One acre-ft per year (893 gpd; 3.4 m³pd). • Three acre-ft per year for multiple households (2,678 gpd; 10 m³pd). • Before 2009, any use less than 2,500 feet in depth.
Oregon	Yes	• Domestic use less than 15,000 gpd (57 m³pd). • Lawn and garden irrigation less than 0.5 acre. • Stockwatering. • Commercial use less than 5,000 gpd (19 m³pd). • 'Stacking' of multiple exempt uses permitted. • Dual use for underground storage and recovery permitted. • New wells must file exempt use with Oregon Water Resources Department.

(Continued)

Table 4.2 (Continued)

Western state or province	Exempt use?	Conditions and comments
Utah	No	• Permit required.
Washington	Yes	• Domestic less than 5,000 gpd (19 m³pd). • Lawn and garden less than 0.5 acre. • Stockwatering. • Industrial use less than 5,000 gpd (19 m³pd).
Wyoming	No	• Permit required. • Domestic, livestock, lawn and garden use less than 1 acre exempt from some permitting and adjudication. • Maximum production rate less than 25 gpm (0.9 m³pm).
Alberta	Yes	• Traditional agricultural uses less than 1,651,075 gallons per year (4,523 gpd; 17 m³pd). • Domestic use less than 330,215 gallons per year (904 gpd; 3 m³pd). • Camp water supplies less than less than 330,215 gallons per year (904 gpd; 3 m³pd).
British Columbia	No	• No permit required.
Manitoba	Yes	• Agricultural and irrigation use less than 6,600 gpd (25 m³pd).
Northwest Territories	Yes	• Any use less than 26,417 gpd (100 m³pd).
Saskatchewan	Yes	• Domestic use less than 1,370,860 gallons per year (3,619 gpd; 14 m³pd).
Yukon	Yes	• Any use less than 26,417 gpd (100 m³pd).

beyond Oregon, exempt well use elsewhere in the western United States and Canada ranges from 35 gpd to over 26,000 gpd (0.13 to 57 cubic meters per day [m³pd]). Like many exurban dwellers across the United States and Canada, Measure 37 claimants tacitly assumed that groundwater could be found in the quantities permitted by OWRD and in a usable quality to meet the demand of the 'sagebrush subdivisions'. Yet, the 'water woes' of wells tapping the sediments and of fractured basalt aquifers going dry are commonplace throughout the Willamette River Basin, despite the area's reputation of being rain-soaked (Moody, 2007). Long-term trends in depths to groundwater and perceived decreases in water-well yield from both the sedimentary and basalt aquifers has led to the designation of approximately 12 areas within the Willamette River Basin as groundwater limited or withdrawn from further appropriation for any or all uses (OWRD and DLCD, 2002).

As in most of the western United States, only a few counties in Oregon regulate the location of subdivisions in groundwater-limited areas. In 1983, Lane

Table 4.3 Generalized summary of 'prove-it' policies for water availability to subdivisions and land use; from Strachan (2001) and supplemented by news media (not comprehensive)

State	Concurrency systems and comments
Arizona	• 1973 – Adequate Water Supply Program initiated. • 1982 – Groundwater Management Act creates Active Management Areas. Subdivisions lacking 100-year adequate water supplies can move forward if developers disclose the information to the first buyers. Subsequent buyers don't have to be informed. • 2008 – SB 1575 extends the authority to adopt the water adequacy provisions to counties or communities outside of an Active. Management Area. Cochise County and Clarkdale adopt standards.
California	• 1998 – California Subdivision Map Act. • Subdivisions > 500 lots – sufficient water supply must be available with written verification of sufficient water supply.
Canada	• 2008 – 'Blueprint' calls for dividing Alberta into six regions according to major watersheds. • The regions would be expected to create individual land plans tied to their water limits.
Colorado	• 2007 – Jefferson County passes well testing ordinance. • 2008 – State legislator proposes law requiring developers 'proof' of water (HB House Bill 1141); died in committee.
Florida	• First comprehensive plan in United States in 1970s. • Enforcement relies on good faith of local government and citizen action.
Maryland	• Encourages counties to adopt concurrency ordinances rather than by state requirement.
Oregon	• 1983 – Lane County Code on Subdivisions delineates quantity and quality groundwater-limited areas. • 1998 – Marion County Sensitive Groundwater Overlay Zones Studies. • 2007 – Benton County ordinance 'prove-it' policy. • 2007 – Clackamas County exploring Benton County model. • 2007 – Measure 49 limits number of exempt wells in state-designated groundwater limited areas.
Texas	• 2007 – Court orders developer to prove groundwater supplies in Weatherford. • 2011 – Parker County adopts subdivision rules and regulations.
Utah	• State Water System Design Standards exist but not enforced. • 2001 – Summit County adopts Concurrency Ordinance and Water Bank for water providers.
Vermont	• Permit for development controlled by regional commission. • Lacks flexibility.

(Continued)

Table 4.3 (Continued)

State	Concurrency systems and comments
Washington	• 2006 – Skagit River Basin plan permits 15.5 cfs of new 'consumptive' exempt use. Basin closed to new exempt use once threshold reached. • 2007 – Citizens group petitions state for moratorium on new exempt wells in Kittitas County. • 2013 – Growth Management Hearings Board for eastern Washington, and a ruling by the state Supreme Court, require Kittitas County to negotiate countywide groundwater well restrictions.
Wyoming	• 1999 – Subdivision law requires proof of water and wastewater. • 2001 – 'Repealed' by watered-down version. • 2008 – Reconsidered by legislature.

County addressed subdivisions in the area near Eugene, located in the southern Willamette River Basin, and identified groundwater-limited areas unsuitable for development based on poor water quality and limited water quantity. The Marion County Rural Zoning Ordinance was implemented to meet the groundwater resources goals and policies of the Environmental Quality and Natural Resources section of the Marion County Comprehensive Plan. The lot density within a Sensitive Groundwater Overlay (SGO) Zone triggers a decision on whether or not a hydrogeologic review is required. In 2007, Benton County passed an ordinance that uses short-term pumping tests to 'prove up' the conclusions made in hydrogeologic studies for subdivision developments relying on exempt wells (see Table 4.3).

A hydrogeologic review provides a background of the general geology and hydrogeology of an area using previous investigations; reviews the hydrogeology with the study area using well logs archived in a database maintained by the OWRD to determine the type of aquifer (e.g., sediment, basalt or alluvium); researches well deepening and replacements and calculates an estimate of the groundwater budget. Additional development is permitted without more extensive and expensive hydrogeologic studies if the existing information has determined that less than 90% of the recharge within the study area would be used after the proposed development is completed, if the proposed development will not adversely affect the long-term water supply of existing and potential new uses on existing parcels and if the additional proposed use will not deplete the groundwater resource over the short or long term. As part of the hydrogeologic review process, the Marion County Planning Division retains experts to conduct 'peer reviews' of the reviews and studies prepared by hydrogeologic experts on behalf of their clients, the applicants for the subdivision.

Petitions for compensation often require the services of land appraisers, resource economists, traffic engineers and water experts. Multiple approaches are inherent to completing the required supplemental economic and technical appraisals.

For example, Jaeger and Plantinga (2007) indicate that two approaches typically used for evaluating the economics of land value result in dramatically different financial results for Measure 37 claims. Hydrogeologic experts representing the first development claim under Measure 37 in Marion County submitted their review on March 12, 2007. In Marion County, approvals of subdivisions can be appealed by outside parties, such as neighboring properties. The appellant can provide hydrogeologic data and studies prepared by other hydrogeologic experts to support the appeal. Concerned citizens organized under the nongovernmental organization (NGO) Keep Our Water Safe Committee (KOWSC) in Salem submitted their hydrogeologic consultant's report and review comments on January 2, 2007. The peer review by the hydrogeologic consultant retained by the Marion County Planning Division was completed on May 3, 2007, and it disagreed with the review by the Measure 37 claimant's consultants. The Marion County Planning and Zoning Commission voted to approve the Measure 37 claim in June 2007, despite the findings by their own consultant. The vote underscores a common problem where '[t]he findings of water science play a role in the politics of allocation and management. But they will generally be subordinate to politics. After all, water flows uphill to money and power' (Stockholm Water Prize Winner Professor John Anthony Allan, as quoted by Bleckner, 2008, p9). According to Ozawa (2005), part of the problem is that the scientific enterprise 'in its purest form' is incompatible with the practical demands of public decision making. How science subordinates to politics is a subject of rich debate and is beyond the scope of this chapter; however, Adler and others (2000), Ozawa (2005) and Pielke (2007) provide excellent introductions to this topic for the interested reader.

Conflict cartography

The spaghetti western water war in Oregon serves as an excellent example of complex environmental conflict resolution (ECR). Emerson, Nabatchi, O'Leary, and Stephens(2003, p3) define ECR as 'complex intergovernmental relationships and myriad disputes involving the environment'. Figure 4.1 provides a situation map or the 'conflict cartography' of the primary stakeholders in the Measure 37 and groundwater development debate using the method outlined by Daniels and Walker (2001) and Papadopoulos (2004). Both Daniels and Walker (2001) and Papadopoulos (2004) underscore the usefulness of graphically displaying the complexity of a situation. Situation maps are an 'illustration, facilitation, and information management approach designed to support the efficient and effective resolution of complex, multistakeholder conflicts' Papadopoulos (2004, p12).

Mapping the political landscape for experts working in groundwater planning for exempt wells is complex. A careful examination of the situation map depicted in Figure 4.1 reveals four overlapping audiences involved in the situation: (1) elected representatives; (2) technical and scientific 'water experts', or the 'meatballs' in the spaghetti bowl; (3) managerial staff and (4) the public at large. Renevier and Henderson (2002) suggest that the interests and options in water conflicts are not easily defined without the assistance of specialists who can interpret causal

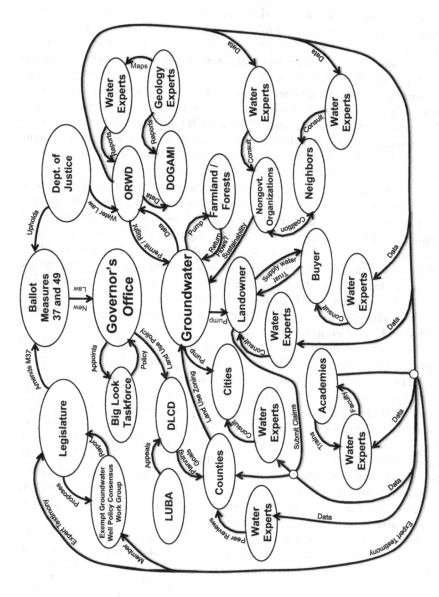

Figure 4.1 Exempt well situation map

chains. Thus, science remains at the core of water issues. The 'types' of science that are involved in this situation are manifold: regulatory or policy science where research questions are framed by legislators and regulators (OWRD, the Department of Land Conservation and Development [DLCD], county, city); anecdotal, local, popular, or 'traditional knowledge' science defined by Ozawa (2005) as information gained by resource users through experience with that resource over time (neighbors, NGOs, water experts); advocacy science typical of courtroom situations (landowners, water experts, neighbors and neighbors who profess 'not in my backyard' [NIMBY]); and academic science typically undertaken by scientists working at government agencies and universities, where the research questions are framed by scientists and driven by rational analysis and expert judgment (Oregon Department of Geology and Mineral Industries [DOGAMI] and the universities). The common thread to the different types of sciences is that 'water experts' come in many forms, ranging from scientists and engineers with training or experience in groundwater resources to well drillers and water witches.

The first part of this conflict rested with 'science' which according to Ozawa (2005, p195), 'the best science is the science whose meaning is agreed to upon by the participants in a decisionmaking process'. At first glance, it would appear that regulatory science was the driver behind this situation. The Marion County Planning Commission passed an ordinance, the landowner filed a Measure 37 claim in an area designated as 'groundwater limited' and retained a water expert to prepare a hydrogeologic review, which – using readily available information maintained in the archives of the OWRD and mapping available from DOGAMI – estimated that the recharge rate to the aquifer(s) underlying the Measure 37 claim complied with the Marion County ordinance.

The KOWSC retained another hydrogeologic expert who challenged the hydrogeologic conceptual model developed by the Measure 37 claimant's hydrogeologist, as well as the water demand calculations, which did not use the 15,000 gpd limit imposed by the OWRD for exempt wells for predicting impacts to nearby springs and lakes. The Measure 37 claimant's hydrogeologist countered the challenge with the notion that using 15,000 gpd for peak allowed exempt use was not reasonable. The law allows for use 'without waste', but the law is silent as to what constitutes 'waste'. Further challenges to the KOWSC's hydrogeologic peer review focused on the conceptual model showing that the contested springs occur at the geologic contact between the basalt and marine sediments; however, there were no data to support this conceptual model that the subject springs were derived from storage in the basalt aquifer.

My examination of the same geologic maps that both experts reviewed indicated yet another conceptual model for the occurrence of the subject springs – that the springs were draining the landslide deposits. At this point in the dispute, the science had evolved from 'regulatory science' associated with compliance with the Marion County ordinance, to 'popular science' associated with debating the quantity of water used by exempt wells, to stealth issue advocacy over the source of water to the springs, sprinkled with academic science regarding an additional potential source of water to the springs.

Complexity

Brogden (2003) defines conflicts as 'complex adaptive systems, which means they are dynamic and nonlinear. In nonlinear systems cause and effect are not always directly linked, proportionate or predictable'. Sword (2006) echoes Brogden's work by stating that '[c]omplexity science deals at a systems level, rather than the constituent parts of systems'. Yet Max-Neef (2005) argues that the logic of linear simplicity seems as strongly grounded as ever, especially in the disciplines related to social action, economics and politics.

Lewicki, Kaufman and others (2003) indicate that water cases are inherently complex. But the groundwater situation is more complex. A high degree of scientific uncertainty exists in groundwater situations due to a lack of field data, professional and ideological disagreements and biased efforts to control the research agenda (Ruhl, 2005). The roles of the technical professionals in the resolution of groundwater resources conflict remain complex and controversial. Access to information, expertise and the quantity and quality of existing data is controversial. Some parties will not accept technical and scientific information. While making a presentation during a local groundwater forum, I heard local water witches and well drillers express suspicions of the information presented by the OWRD scientists and engineers.

But Lewicki, Kaufman and others (2003, p255) point out that the 'science of measuring and managing water . . . is fraught with uncertainty . . . [and] watersheds are complex and dynamic systems that can change quickly and erode quietly with time'. They also indicate that 'water-based conflicts are exacerbated by the fact that the different areas of a watershed can have remarkably different features'. As a consequence, land-use activities, or changes in land-use activities negotiated through collaborative efforts, that have no effect in one part of a watershed can be 'highly detrimental to aquatic life elsewhere'.

Part of the complexity of the Measure 37 situation rests with the well database maintained by the OWRD and used by all of the water experts. According to Leitman (2005), organizing large data sets so they can be understood is not a trivial task. In the Apalachicola-Chattahoochee-Flint basin situation in the southeastern United States, Leitman (2005, p83) reported that 'no consensus was reached on how data should be presented, and each party presented the data as they saw fit leading to a lack of understanding among the negotiating parties'. The abundance of well data that each party used in their analysis was large and some of the debate between experts rested not only on the geology, but also on the well database. For example, part of the review mandated by Marion County focused on well deepenings, well replacements and water level declines. The hydrogeologist for the Measure 37 claimant reported a statistic on the number of basalt wells in the subject area that was below the statistical threshold cited by Marion County as indicative of potential issues with groundwater availability. Yet the peer review by the county's expert reported that additional information provided to Marion County by KOWSC indicated more basalt wells than the number previously reported by the Measure 37 claimant's hydrogeologist. KOWSC's data collection is consistent with a 'citizen science' or community-based effort at environmental data collection and reporting, described as 'bucket brigades' by O'Rourke and Macey (2003).

My role and the role of the Oregon University System in this situation was initially limited to a short meeting that I held with the KOWSC, several presentations that I made to watershed councils, county planning commissions and technical conferences and the testimony I provided to legislative committees. At no time did any of the consultants contact me or other academic experts at state universities for assistance or technical reviews. Considering the complex and technical nature of the groundwater setting, Leitman (2005) found the limited involvement of the expertise of the state universities perplexing in a comparable complex water dispute in the southeastern United States. Susskind (2005) and Ritchey (2013) indicate that learning about complex and 'wicked' systems can only occur through purposeful experimentation – one of the many missions of the academies. The apparent lack of use of the academic water experts may be due to the fact that 'each group generally sees their own interpretation of the physical reality as science-based: they charge the other sides with disregarding science and being reckless or fear-driven, respectively. There is also a large symbolic component to the debate, framed as a battle between the primacy of economics and growth and the primacy of the precautionary principle' (see Roeder, 2005, p115).

Dueling experts syndrome

According to Wade (2004), common causes of conflict focus on missing information, inaccurate data and procedures of data analysis. Wade (2004, p420) suggests that data conflicts can be settled by 'the dispassionate opinions of one or more alleged experts about history, causation or the future'. However, the 'darker' side of using experts to settle conflicts rests with 'helping' disputants 'solve' their problem, and the conflict escalates because the expert(s) become part of the problem instead of part of the solution. Glennon (2006) has referred to some of these experts as 'hydrostitutes'.

KOWSC reported that they had to fight an unethical and flawed hydrogeologic review prepared by a hydrogeologist who was previously involved in some of the assessments completed in the neighboring areas. Apparently, the shoddy work led the Oregon State Board of Geologists Examiners to take action against the hydrogeologist. This historical relationship with a 'water expert' led KOWSC to suspect all water experts as 'reckless' and to identify their opinions with that of the water expert's client.

Wade (2004) describes the disputant's behavior in the dueling experts syndrome as having the following characteristics:

- Employs an expert ('ours is the best in the field')
- Hires an expert who has reputation for favoring disputant's preferred outcome ('reputational partiality')
- Tells different stories to own expert ('garbage in, garbage out')
- Makes expressed or implied hints at the advice he or she wants from the expert

Dueling experts apparently exhibit the following characteristics:

- Do not consult with each other ('delusionary isolation')
- Tell client what they want to hear ('you get what you pay for')
- Refrain from providing 'best to worst alternatives' ('delusionary certainty')
- Long, incoherent reports ('mysterious complexity')
- Refuse to share draft reports ('no early doubts of compromises')

The Marion County Measure 37 claim situation had many of the attributes of the dueling experts syndrome. The Measure 37 claimant retained private consultants apparently because of their reputation as skilled hydrogeologists, but may have also retained them because they also provided consulting services to KOWSC, who challenged the work completed by the 'water expert' that was scrutinized by the state regulatory agency.

The hydrogeologic review completed for the Measure 37 claimant closely fits the description of a long, incoherent report, thus adding to the 'mysterious complexity' of the groundwater situation. The claimant's expert report was over 400 pages in length; copies of well logs derived from the OWRD database comprise the bulk of the report. Likewise, the peer review completed by the county's expert suggested that the claimant's expert report refrained from providing 'best to worst alternatives' in reporting the water balance calculations, the water use calculations, the well-deepening replacements and water level declines in nearby wells and in presenting a conceptual model of the local hydrogeology that favored the development of the proposed Measure 37 subdivisions.

It is clear that when a dispute evolves into the dueling experts syndrome, parties in water management disputes pursue their conflicting interests and overlook common interests (Quirk, 2005). Sometimes the lack of respect afforded to and between technical professionals working on water-related issues also leads to problems in negotiating a water use dispute (Fitzhugh and Dozier 2001). And the dueling experts syndrome is good business for conflict beneficiaries, whereas trying to resolve a dispute through collaboration may save money. For example, in a study of hundreds of lawsuits, Glater (2008) reports that 'the vast majority of cases do settle – from 80 to 92 percent by some estimates. On average, getting it wrong cost plaintiffs at about $43,000; the total could be more because information on legal costs was not available in every case. For defendants, who were less often wrong about going to trial, the cost was much greater: $1.1 million'.

Applying the Hydro-Trifecta Framework to the spaghetti western water war

Rural development has historically relied upon wells. The perception that groundwater can be found everywhere and recovered in quantities and at a quality fit for human consumption and small scale irrigation is a myth. Yet that is the tacit assumption that low-quantity well users make when developing subdivisions and farming in rural areas.

The exempt well issue results in many individual users with little to no government planning and control. The spectacular increase in groundwater use has been mainly driven by economic reasons: the full direct cost of groundwater use is usually a small fraction of the value of crops obtained with groundwater abstraction; likewise, groundwater use in rural subdivisions development is also a small fraction of the value of the housing dependent on groundwater abstraction. With the silent revolution come problems such as the degradation of groundwater quality, excessive drawdown of groundwater levels, land subsidence, reduction in spring flow and base flow and groundwater-dependent ecosystems. But the silent revolution also leads to a strengthened capacity of groundwater beneficiaries, such as real estate developers, well drillers and landowners, to form political lobbies with political clout. Yet the apparent disconnect between land use and water use is building surprises and consigning public priorities to resources to a death by a thousand cuts (Van de Wetering, 2007).

Scale targeted skills

Exempt wells constitute a somewhat unique situation, with conflicts occurring within and between multiple scales. We see the classic interpersonal conflict between the dueling experts. Intersectoral conflicts exist between land developers and landowners. Interagency conflicts exist between water resources managers and land conservation managers and between county and state agencies. Exempt well requirements clearly vary from state to state and province to province and little has been written about these potential interstate conflicts. Likewise, neighboring states and provinces share little in the way of treaties or agreements regarding exempt wells, thus setting the stage for future international conflicts.

Competency targeted skills

On the basis of the existing information, it is clear that the various parties have a high level of knowledge with adept functional competencies in report writing and technical presentations. But there is a wide disparity in competency in the field of hydrogeology, with respect to what constitutes hydromyths. The discourse between the different experts indicated a broad spectrum of values regarding professionalism. Cultural differences between urban and rural lifestyles and the associated NIMBY syndrome indicate a dominant issue area for negotiations.

Program targeted skills

Dueling experts situation

Wade (2004) describes two approaches to responding to the dueling experts syndrome: (1) reframing the expert conflict into a problem that can be addressed cooperatively and (2) systematically identifying options for resolving the expert dilemma. In many regards, Wade's suggested approach sounds like the collaborative learning approach used on other environmental conflicts, where disputants can learn from each other (Daniels and Walker, 2001, 2012).

To begin the process of resolving the vexing question of 'What can be done about the current differing views of the experts', Wade (2004) proposes the following 12-step approach:

1 Joint meetings between disputants and their dueling experts to 'try to convince' by each party pointing out the strengths and weaknesses of experts.
2 Joint presentation of experts on why differences exist through a mediator.
3 Experts answer list of written questions.
4 Experts prepare jointly signed explanation.
5 Outside advisory expert(s) attends mediation or negotiation.
6 Outside expert(s) writes a nonbinding opinion.
7 Outside expert(s) writes a binding decision.
8 Mediator creates doubt by introducing new or hypothetical facts ('what ifs')
9 Split the difference.
10 Trade chips.
11 Toss a coin.
12 Refer decision to a judge.

While Leach and Sabatier (2003, p167) indicate that it is better for facilitators to be disinterested, they also suggest that non-disinterested facilitators can be effective facilitators if they follow through on an in-kind basis or as volunteers because paid facilitators 'may evoke feelings of resentment among those whose affairs are being facilitated for financial gain', and that '[w]atershed stakeholders may also be skeptical of overly polished or excessively managed processes'. Scher (1997) suggests that it is wise to seek out advisors such as scientists or engineers with process skills and technical expertise to assist with complex environmental disputes. In the case of the dueling experts and Measure 37 claims, neutral boards already exist, such as the Oregon Board of Geologists Examiners, or technical experts could be derived from the Oregon University System.

Rather than attempting to reach agreement on the many issues facing the dueling experts situation, Adler (2000) suggests that it is more important to get to 'maybe' rather than to 'yes' when discussing the upsides and downsides to all potential solutions, and that once 'maybe' is reached, then what follows is mutually focused thinking and productive talk. While each party will more than likely be oriented toward their own self-interest, the 'sequencing' approach to group facilitation, as described by Kaner (1996), might work well with the dueling experts situation. The topic of the sequencing might focus on one topic by the neutral board(s), such as 'What does safe yield mean to you?' or 'What does sustainability mean to you in terms of groundwater resources?'

In the case of disputes over groundwater, a common practice is to have the experts prepare a jointly signed explanation. As discussed in earlier chapters, the Great Basin Aquifer underlying the states of Utah, Nevada, Idaho and Oregon is targeted for development by the Southern Nevada Water Authority in order to provide groundwater supplies approaching 0.21 km^3 per year to Las Vegas. While Bredehoeft and Durbin (2009) acted as consultants to opposing sides of the groundwater development project, both parties prepared a peer-reviewed technical paper describing the groundwater modeling predictions. However, linking technical dialogue between the dueling experts to broader public learning will more than likely remain a challenge (Roeder, 2005).

Spaghetti western water wars

Using the Progress Triangle Framework of Walker and Daniels (2003) to assess the collaborative potential of the conflict over private property development, well interference, conceptual models of the local hydrogeology, the 'safe yield' of the target aquifers and the anticipated water use by the 'exempt' wells revealed that the lowest ratings and rankings did not match well – the situations with the perceived need for collaboration and with the greater collaborative potential included the development of private property, followed by quantification of the exempt well use, and the 'safe yield' calculations. The situations with lower potential for collaboration included the conceptual models of the local hydrogeology and the perceived interference between the existing the new wells associated with the private property development.

The apparent decrease in collaborative potential over the conceptual models and well interference issues may be due to how the water experts 'identify' or take ownership of their models. Both the conceptual models and well interference issues are linked – the well interference cannot be predicted with any degree of confidence without the conceptual model. And it is in the best interests of both the Measure 37 claimant and the NGOs to develop conceptual models that either do or do not show well interference. Regardless of whether the conflict initially appears to be an interest or resource-based conflict, it is important to acknowledge that identity is one of the foundations for nearly all conflicts (Rothman, 1997). Rothman indicates that many conflicts are poorly diagnosed, since identity conflicts are usually misrepresented as disputes over tangible resources. While collaborative learning and systems thinking are valuable tools in assessing program-targeted skills for addressing wicked problems, such as exempt wells and spaghetti western water wars, Ritchey's (2013) summary of the literature on general morphological analysis indicates that group-facilitated discussions can also work to (1) accommodate multiple alternative perspectives, rather than prescribe single solutions, and (2) function through group interaction and iteration, rather than back-office calculations.

In 2008, the state legislature convened an Exempt Groundwater Well Policy Consensus Work Group through the Oregon Consensus Program at Portland State University to further explore the issue of exempt wells. And while the work group was composed of state legislators, state agencies, nongovernmental groups, many business-related organizations, universities and well drilling organizations, little was achieved in terms of reaching consensus, in part due to the perception that overall groundwater use by exempt wells was estimated at 7% of total

Figure 4.2 Exempt well Circle of Conflict; from Vinett and Jarvis (2012)

groundwater withdrawals in Oregon. Many in the work group also expressed frustration that much of the time spent in the work groups was dedicated to educating the seasoned facilitator on the hydrogeologic and well-drilling jargon.

Likewise, the Exempt Wells conference held in 2011 provided a unique opportunity to expand learning across all scales. Vinett (2011) compiled a mega-situation map of the conference that was extremely complicated, so much so that Vinett and Jarvis (2012) attempted to reduce the complexity of the map to a Circle of Conflict as a means to inventory the triggers of conflicts, as depicted in Figure 4.2. Reducing the situation map to this simplified form indicated that nearly every characteristic of an intractable conflict, as defined by Campbell (2003), was exhibited by the spaghetti western water war.

Which siting mechanism should be used?

Water security?

The spaghetti western water war and the associated dueling experts situation place groundwater quantity and quality at risk across large areas and also risk interfering with surface water resources. It could be reasonably argued that the risk extends to the balancing between natural 'security resources' (food and water) and equitability between the individuals, communities (subdivisions, counties, states and provinces) and nations involved (Zeitoun, 2011).

Water diplomacy?

This siting mechanism acknowledges that water crosses multiple domains and multiparty negotiation and coalitional behavior. It primarily addresses interests and sets its sights on the flexible uses of water and joint fact finding to create value.

While the top-down administrative decision making typical of government agencies pushes groups to accentuate their differences, rather than search for common ground, as described by Wondolleck and Yaffee (2000), the incentives for all parties to avoid litigation will not solve the problems of competition over limited natural and financial resources. Conversely, the lack of readily available funding for data collection to supplement both technical and public learning also may serve as an incentive for participants to complete some of the required data gathering on their own. For example, O'Rourke and Macey (2003) describe how 'bucket brigades' fill in the gap between monitoring and data collection for enforcement activities in neighborhoods surrounding industrial facilities. While O'Rourke and Macey (2003) report that the bucket brigades are instigated by community members and facilitated and intermediated by NGOs, they have been only partially accepted by government agencies.

Susskind (2005) and Islam and Susskind (2013) indicate that joint fact finding (JFF), such as by a bucket brigade, encourages stakeholders to specify the information that they desire to collect by expanding the technical capacity of participants through facilitated dialogue. Ozawa (2005) recommends JFF because it acknowledges 'surprises' and forces resource managers or elected decision makers to issue public statements clarifying the basis for the discrepancies when analyses or reports point to opposing policy prescriptions. Public confusion can discredit decision makers and lead to the rejection of all scientific work and a retreat to other forms of decision making. Likewise, periodic review and reassessment of monitoring and data collection are essential and eminently practical, given that industries, firms, private individuals and municipalities are sensitive to capital investment costs and rely on a certain degree of stability and predictability in decisions (Ozawa, 2005). Other incentives to the JFF approach include the political risk of ignoring consensual proposals and data collection efforts – any elected body that ignores the consensual proposals of a truly representative stakeholder group that has done its homework does so at substantial political risk (Susskind, 2005).

Water conflict transformation?

Conflicts and disputes arise about groundwater allocation and protection and use because it's personal – people are sustained by their water (Vinett and Jarvis, 2012). At first glance, it would seem that the core motive influencing decision making under this siting mechanism comes from institutions. But van Vugt (2009) indicates that identity works toward action by connecting groups of competitors to move toward action. The number of exempt wells demands thinking beyond jurisdictional boundaries, beyond basin boundaries and beyond just groundwater to ways of 'blurring group boundaries' by referencing aquifer communities and implying that we are all in this together, as suggested by Shah (2009).

The graduated circle of transdisciplinarity

What exists?

The list of what exists is long and often contradictory. What is known is that millions of wells exist across the globe, some in use and many not. The plethora of different regulations, or lack thereof, governing these wells indicates an extremely wicked problem that will be difficult to tackle through any imagination, much less the transdisciplinary imagination. There are about as many different opinions on the impact as well as the value of exempt wells as there are exempt and illegal wells.

What are we capable of doing?

The current model regarding exempt wells and the use of expertise to resolve water conflict fits well with the Daniels and Walker (2001) model of 'Tech-Reg' as a source of conflict, but the institutions could change their model to be more collaborative and provide more of a leadership role. Jones (2005) indicates that there is a powerful and lasting legacy of leadership based on the command-and-control leadership (Tech-Reg) model. Learning a new leadership language that better fits the complex and adaptive world of water resources and the role of public learning in these initiatives might be better served by 'leaders' with the following traits:

- Collaborative leaders who inspire political and personal commitment and action
- Collaborative leaders who function as peer problem solvers
- Collaborative leaders who build broad-based involvement in the collaborative enterprise
- Collaborative leaders who work to sustain hope and encourage participation in the consensus building process

Wondolleck and Yaffee (2000) argue that successful collaborative efforts have one or two individuals who 'lead by example'; sometimes the leader is a forward-looking agency official, sometimes the leader is a highly respected expert. Leach and Sabatier (2003) also suggest that local leadership is important; given that water issues rarely fit within political boundaries, they counsel that the level of agreement reached can be enhanced if the leadership can increase the participation from government officials who hold permitting authority to guide 'what we want to do', the technical expertise and the funding that continue collaborative efforts and project implementation.

What do we want to do?

When it came to the Oregon situation, consensus was reached on two items: (1) recommending additional funding for state agencies to collect data and conduct studies for groundwater resource management and (2) recommending that state agencies be provided specific exempt well locations (see Oregon Consensus, 2013).

Recognizing that changes in regulation or attempts to mitigate impacts of exempt wells would likely be met with social or political resistance, Bracken (2010) and Vinett (2011) briefly discuss options for collaborative or negotiated governance to reach feasible agreements. Despite differing concerns and dissimilar opinions on next steps and appropriate actions, the following list of potential agreements were compiled during the international Exempt Wells (2011) conference:

- Scalpels, as opposed to hammers, are probably the best weapon to be used in the 'war', as described by Bracken (2012)
- Basin-specific, local collaboration is recommended, especially for transboundary waters (both interstate and international)
- Aim for long-term leadership, political will and funding dedicated to ending the 'war' as summarized by Vinett and Jarvis (2012)
- Decisions are based on land use and population control
- Scope of permit exemptions still needs to be determined
- Addressing whether water from exempt wells is a 'property right', a 'human right' or a 'right to life' as defined by the U.S. Constitution, or all simultaneously as summarized by Vinett and Jarvis (2012)
- More groundwater and aquifer data including long-term analysis of these data
- Educate, educate, educate as recommended by Richardson (2012)
- Consider pilot projects as suggested by Embleton (2012) and Ziemer and others (2012)
- Recognize the importance of relationships, especially developing in local solutions

Washington State, like many western states and some Canadian provinces, is taking legislative action to prevent the misuse of the infamous 'exempt well loophole' (Moran, 2013). However, even with the identification of this solution, there is a recognized risk associated with ending the spaghetti western water war, and that is the power of the status quo, as posited by Hamman (2005, p129): 'Those individuals, communities, and institutions that benefit from the current allocation or perceive they will suffer from a change have great power to defend the status quo'.

What we must do?

Until the political melodrama on exempt and 'illegal' wells plays out, and given the existing situation, with millions of wells across the world constructed under many different institutional constructs, coupled with the power of the status quo, perhaps it is best to consider the spaghetti western water war intractable and unwinnable. According to Campbell's (2003) summary of the characteristics of intractability, the factors that make the exempt well situation particularly intractable include (1) the fundamental value differences between parties; (2) the fact that conflicts have persisted for many years; (3) the power imbalances, particularly the urban-rural divides, prevalent between the disputing parties; (4) the strongly held beliefs regarding the many types of 'rights' associated with exempt wells; (5) the many interlocking issues and, most importantly; (6) the threats to parties' individual or collective identities.

Considering the fact that these wells are installed in many different hydrogeologic settings using many different drilling techniques and well materials, and often in closely spaced settings, and that they have rare follow-up inspection and servicing to assess well seal integrity and the corrosion of well materials, it is in the best interest of all to globally address end-of-life well stewardship. The financial premium associated with a 'death by a thousand wells' to aquifers storing high-quality water is incalculable and it is something that we all must do something about.

Universal well-care programs exist in some states and provinces through the establishment of well-abandonment reimbursement funds. Experts postulate that it can be assumed that most exempt wells are not plugged when they are abandoned and may serve as conduits for the movement of contaminated water into an aquifer. The risk of such contamination is likely to increase after 35 years, as the well casing rusts away or collapses or land use changes (Bracken, 2010; Jarvis and Stebbins, 2012). Inventories of abandoned wells by provinces, states and counties should become 'shovel-ready' projects that qualify for federal, state or provincial and local funds.

References

Adler, P.S. (2000) 'Water, science, and the search for common ground', Mediate.com, www.mediate.com/articles/adler.cfm

Adler, P.S., Barrett, R.C., Bean, M.C., Ozawa, C.P. and Rudin, E.B. (2000) 'Managing scientific and technical information in environmental cases: Principles and practices for mediators and facilitators', Mediate.com, www.mediate.com/articles/pdf/envir_wjc1.pdf

Anderson, E.W. (1999) 'Geopolitics: International boundaries as fighting places', in C.S. Gray and G. Sloan (eds) *Geopolitics, Geography, and Strategy*, Frank Cass, Portland, OR, pp125–136

Applegate, G. (2002) *The complete book of dowsing: The definitive guide to finding underground water*, Element Books Ltd, Shaftesbury, UK

Bird, C. (1993) The divining hand: The art of searching for water, oil, minerals, and other natural resources or anything lost, missing, or badly needed, Whitford Press, Atglen, PA

Birkenholtz, T. (2008a) 'Contesting expertise: The politics of environmental knowledge in northern Indian groundwater practices', Geoforum, vol 39, pp466–482

Birkenholtz, T. (2008b) 'Groundwater governmentality: Hegemony and technologies of resistance in Rajasthan's (India) groundwater governance', *The Geographical Journal*, vol 175, no 3, pp208–220

Bleckner, S. (2008) 'Do the right thing a little badly', Stockholm Water Front, 2 July, pp8–9

Bracken, N. (2010) 'Exempt well issues in the West', *Environmental Law*, vol 40, pp141–253

Bracken, N. S. (2012) 'Scalpels v. hammers: Mitigating exempt well impacts: universities council on water resources', *Journal of Contemporary Water Research & Education*, no 148, pp24–32

Bredehoeft, J. and Durbin, T. (2009) 'Ground water development – the time to full capture problem', *Groundwater*, vol 47, pp506–514

Brogden, M. (2003) 'The assessment of environmental outcomes', in R. O'Leary and L. B. Bingham (eds) *The Promise and Performance of Environmental Conflict Resolution*, Resources for the Future, Washington, DC, pp277–300

Bryner, G. (2004) 'Western United States groundwater law', *The Water Report*, 15 July, pp9–16

Bryner, G. and Purcell, E. (2003) *Groundwater law sourcebook of the western United States*, Natural Resources Law Center, University of Colorado at Boulder, Boulder, CO

Cameron, A. B. (2009) 'A study in transboundary ground water dispute resolution', *Sea Grant Law and Policy Journal*, 2009 Symposium, Water Quantity: Ongoing Problems and Emerging Solutions, http://nsglc.olemiss.edu?SGLPJ?presentations_09/cameron.pdf

Campbell, M. C. (2003) 'Intractable conflict,' in R. O'Leary and L. B. Bingham (eds) *The Promise and Performance of Environmental Conflict Resolution*, Resources for the Future, Washington, DC, pp90–110

Chevalking, S., Knoop, L. and Van Steenbergen, F. (2008) *Ideas for groundwater management*, MetaMeta and IUCN, Wageningen, the Netherlands

Daniels, S. E. and Walker, G. B. (2001) *Working through environmental conflict: The collaborative learning approach*, Prager, Westport, CT

Daniels, S. E. and Walker, G. B. (2012) 'Lessons from the trenches: Twenty years of using systems thinking in natural resource, conflict situations', *Systems Research and Behavioral Science*, vol 29, pp104–115 doi:10.1002/sres.2100

Deming, D. (2002) 'Water witching and dowsing', *Groundwater*, vol 40, no 4, pp 450–452

Ellis, A. J. (1917) 'The divining rod: A history of water witching', U.S. Geological Survey Water-Supply Paper 416

Embleton, D. G. (2012) 'Use of exempt wells as natural underground storage and recovery systems', *Journal of Contemporary Water Research & Education*, no 148, pp44–54

Emerson, K., Nabatchi, T., O'Leary, R. and Stephens, J. (2003) 'The challenges of environmental conflict resolution', in R. O'Leary and L. B. Bingham (eds) *The Promise and Performance of Environmental Conflict Resolution*, Resources for the Future, Washington, DC, pp3–26

Exempt Wells (2011) 'Exempt Wells: Problems and Approaches in the Northwest', Proceedings of the May 17–18, 2011, Conference at WallaWalla, WA, Washington State University, Pullman, WA, http://cm.wsu.edu/ehome/index.php?eventid=24592&)

Fitzhugh, J. H. and Dozier, D. P. (2001) 'Finding the common good: Sugarbush water withdrawal', Mediate.com, www.mediate.com/articles/dozier.cfm

Glater, J. D. (2008) 'Study finds settling is better than going to trial', *The New York Times*, 7 August, www.nytimes.com/2008/08/08/business/08law.html?_r = 0

Gleick, P. (2009) 'Water Brief No. 4: Water conflict chronology', in P. Gleick (ed) *The World's Water 2008–2009*, Island Press, Washington, DC, pp151–193

Gleick, P. H. and Heberger, M. (2012) 'Water conflict chronology', in *The World's Water: The Biennial Report on Freshwater Resources*, vol 7, Island Press, Washington, DC, pp175–205

Glennon, R. (2002) *Water follies: Ground water pumping and the fate of America's fresh waters*, Island Press, Washington, DC

Glennon, R. (2006) 'Tales of french fries and bottled water: The environmental consequences of groundwater pumping', *Environmental Law*, vol 37, no 3, pp3–13

Hamman, R. (2005) 'The power of the status quo', in J. T. Scholz and B. Stiftel (eds) *Adaptive Governance and Water Conflict: New Institutions for Collaborative Planning*, Resources for the Future, Washington, DC, pp125–129

Institute for Water and Watersheds (2013) 'Related documents: Summary documents about the roundtables', www.water.oregonstate.edu/roundtables/docs.htm

Islam, S. and Susskind, L. E. (2013) *Water diplomacy: A negotiated approach to managing complex water networks*, RFF Press, Routledge, New York, NY

Jaeger, W. K. and Plantinga, A. J. (2007) 'The economics behind Measure 37', Oregon State University Extension Service, EM 8925

Jarvis, W. T. and Stebbins, A. (2012) 'Examining exempt wells: Care for exempt wells provides opportunities for the water well industry', *Water Well Journal*, September 2012, pp23–27

Jones, R. M. (2005) 'Leadership and public learning', in J. T. Scholz and B. Stiftel (eds) *Adaptive Governance and Water Conflict: New Institutions for Collaborative Planning*, Resources for the Future, Washington, DC, pp164–173

Kaner, S. (1996) *Facilitator's guide to participatory decision-making*, New Society Publishers, Gabriola Island, BC

Knapp, G. and Nelson, A. C. (1993) *The regulated landscape*, Lincoln Institute of Land Policy, Cambridge, MA

Leach, W. and Sabatier, P. (2003) 'Facilitators, coordinators, and outcomes', in R. O'Leary and L. B. Bingham (eds) *The Promise and Performance of Environmental Conflict Resolution*, Resources for the Future, Washington, DC, pp148–171

Leitman, S. (2005) 'Apalachicola-Chattahoochee-Flint Basin: Tri-State negotiations of a water allocation formula', in J. T. Scholz and B. Stiftel (eds) *Adaptive Governance and Water Conflict: New Institutions for Collaborative Planning*, Resources for the Future, Washington, DC, pp74–88

Lewicki, R. J., Kaufman, S., Wiethoff, C. and Davis, C. B. (2003) 'Comparing water cases', in R. J. Lewicki, B. Gary and M. Elliot (eds) *Making Sense of Intractable Environmental Conflicts: Concepts and Cases*, Island Press, Washington, DC, pp255–271

Llamas, M. R. and Martínez -Santos, P. (2005a) 'The silent revolution of intensive ground water use: pros and cons', *Groundwater*, vol 43, no 2, p161

Llamas, M. R. and Martínez-Santos, P. (2005b) 'Intensive groundwater use: Silent revolution and potential source of social conflicts', *American Society of Civil Engineers Journal of Water Resources Planning and Management*, vol 131, no 4, pp337–341

Max-Neef, M. A. (2005) 'Foundations of transdisciplinarity', *Ecological Economics*, vol 53, pp5–16

Moench, M. (2004) 'Groundwater: The challenge of monitoring and management', in P. Gleick (ed) *The World's Water 2004–2005*, Island Press, Washington, DC, pp79–100

Moody, J. (2007) 'Water woes: Rain or not, wells run dry', Albany County-Herald, 7 April

Moran, K. A. (2013) 'The role of policy and collaboration in decommissioning exempt water wells in Washington State', unpublished Master of Science thesis in Water Resources Policy and Management, Oregon State University, Corvallis, OR

Mortenson, E. (2008) 'Measure 49 scales back development', *The Oregonian*, 20 June, http://blog.oregonlive.com/breakingnews/2008/06/measure_49_will_scale_back_rur.html

National Ground Water Association (2010) 'Groundwater facts', www.ngwa.org/Fundamentals/use/Pages/Groundwater-facts.aspx

Nowlan, L. (2005) 'Buried treasure: Groundwater permitting and pricing in Canada', with Case Studies by Geological Survey of Canada, West Coast Environmental Law, and

Sierra Legal Defense Fund, www.waterlution.org/sites/default/files/gw_permitting_pricing_canada_e.pdf

O'Rourke, D. and Macey, G. P. (2003) 'Community environmental policing: Assessing new strategies of public participation in environmental regulation', *Journal of Policy Analysis and Management*, vol 22, no 3, pp383–414

Oregon Consensus (2013) Exempt Groundwater Well Policy Consensus Group, http://oregonconsensus.org/projects/441/

Oregon Water Resources Department (OWRD) and Oregon Department of Land Conservation and Development (DLCD) (2002) 'Ground water supplies in the Willamette Basin', www.oregon.gov/owrd/gw/docs/rptgwsupplieswmbasin10-2002.pdf

Ozawa, C. (2005) 'Putting science in its place', in J. T. Scholz and B. Stiftel (eds) *Adaptive Governance and Water Conflict: New Institutions for Collaborative Planning*, Resources for the Future, Washington, DC, pp185–195

Papadopoulos, N. (2004) 'Conflict cartography: A methodology designed to support the efficient and effective resolution of complex, multistakeholder conflicts', ViewCraft, http://compendiuminstitute.net/compendium/papers/conflictcartography42.03.pdf

Peterson del Mar, D. (2003) *Oregon's promise: An interpretive history*, Oregon State University Press, Corvallis

Pielke, R. A. (2007) *The honest broker: Making sense of science in policy and politics*, Cambridge University Press, Cambridge, UK

Portland State University (2013) 'Measure 37 database', www.pdx.edu/ims/measure-37-database

Pruesch, M. (2008) 'Oregonians push for ideas to create rural prosperity', *The Oregonian*, 22 August

Putnam, L. L. and Peterson, T. (2003) 'The Edwards Aquifer dispute: Shifting frames in a protracted conflict', in R. J. Lewicki, B. Gray and M. Elliot (eds) *Making Sense of Intractable Environmental Conflicts*, Island Press, Washington, DC, pp127–158

Quirk, P. J. (2005) 'Restructuring state institutions' in J.T. Scholz and B. Stiftel (eds) *Adaptive Governance and Water Conflict: New Institutions for Collaborative Planning*, Resources for the Future, Washington, DC, pp204–212

Renevier, L. and M. Henderson (2002) 'Science and scientists in international environmental negotiations', in L. Susskind, W. Moomaw and K. Gallagher (eds) *Transboundary Environmental Negotiation – New Approaches to Global Cooperation*, Jossey-Bass, A Wiley Company, San Francisco, CA, pp107–129

Richardson, J. J., Jr. (2012) 'Existing regulation of exempt wells in the United States', *Universities Council on Water Resources, Journal of Contemporary Water Research & Education*, no 148, pp3–9

Ritchey, T. (2013) 'Wicked problems: Modelling social messes with morphological analysis', *Acta Morphologica Generalis*, vol 2, no 1, pp1–8

Roeder, E. (2005) 'Aquifer storage and recovery: Technology and public learning', in J. T. Scholz and B. Stiftel (eds) *Adaptive Governance and Water Conflict: New Institutions for Collaborative Planning*, Resources for the Future, Washington, DC, pp106–116

Rothman, J. (1997) *Resolving identity-based conflict in nations, organizations and communities*, Jossey-Bass, A Wiley Company, San Francisco, CA

Ruhl, B. S. (2005) 'Representation, scientific learning, and the public interest', in J. T. Scholz and B. Stiftel (eds) *Adaptive Governance and Water Conflict: New Institutions for Collaborative Planning*, Resources for the Future, Washington, DC, pp130–134

Scher, E. (1997) 'Using technical experts in complex environmental disputes', Mediate.com, www.mediate.com/articles/expertsC.cfm

Schwartz, F. W. (2013) 'Folk beliefs and fracking', *Groundwater*, vol 51, no 4, p479

Shah, T. (2009) *Taming the anarchy: Groundwater governance in South Asia*, Resources for the Future, Washington, DC

Stephens, C. E. (1952) 'Witching for water in Oregon', *Western Folklore*, vol 11, no 3, pp204–207

Stone, A. (1999) 'Water dowsing is bad news for groundwater' (editorial), *National Driller's Journal*, www.nationaldriller.com/publications/3/editions/1243

Strachan, A. (2001) 'Concurrency laws: Water as a land-use regulation', *Journal of Land, Resources, and Environmental Law*, vol 21, p435

Susskind, L. (2005) 'Resource planning, dispute resolution, and adaptive governance', in J. T. Scholz and B. Stiftel (eds) *Adaptive Governance and Water Conflict: New Institutions for Collaborative Planning*, Resources for the Future, Washington, DC, pp141–149

Sword, D. (2006) 'Complexity Science Analysis of Conflict', Meidate.com, www.mediate.com/articles/swordL1.cfm

Thomasson, F. (2005) 'Local conflict and water: Addressing conflicts in water projects', Unpublished report prepared for the Swedish Water House, www.swedishwaterhouse.se/swh/resources/20051017114417Conflicts_Water_Projects_050823.pdf

Tidwell, V. C. and van den Brink, C. (2008) 'Cooperative modeling: Linking science, communication, and ground water planning', *Groundwater*, vol 46, no 2, pp174–182

Trout Unlimited (2007) 'Gone to the well once too often: The importance of ground water to rivers in the West', http://old.tu.org/conservation/western-water-project/ground-water

Van de Wetering, S. B. (2007) 'Bridging the governance gap: Strategies to integrate water and land use planning', Collaborative Governance Report No. 2, Public Policy Research Institute, the University of Montana

van Vugt, M. (2009) 'Triumph of the commons: Helping the world to share', *New Scientist*, no 2722, pp40–43

Vinett, M. A. (2011) 'North American groundwater disputes: A study of the impact of exempt wells', MS Terminal Project, Conflict and Dispute Resolution Program, University of Oregon

Vinett, M. and Jarvis, T. (2012) 'Conflicts associated with exempt wells: A spaghetti western water war', *Journal of Contemporary Water Research & Education*, no 148, pp10–16

Vogt, E. Z. and Hyman, R. (1979) *Water witching USA*, University of Chicago Press, Chicago, IL

Wade, J. H. (2004) 'Dueling experts in mediation and negotiation: How to respond when eager expensive entrenched expert egos escalate enmity', *Conflict Resolution Quarterly*, vol 21, no 4, pp419–436

Walker, G. B. and Daniels, S. E. (2003) 'Assessing the promise and potential for collaboration: The Progress Triangle Framework' International Environmental Communication Association, http://theieca.org/contents-2003-conference-communication-and-environment-proceedings

Wax, E. (2006) 'Dying for water in Somalia's drought', *The Washington Post*, 14 April

Wondolleck, J. M. and S. L. Yaffee (2000) *Making collaboration work: Lessons from innovation in natural resource management*, Island Press, Washington, DC

Zeitoun, M. (2011) 'The global web of national water security', *Global Policy*, vol 2, no 3, pp286–296

Ziemer, L., Bates, S., Casey, M. and Montague, A. (2012) 'Mitigating for growth: A blueprint for a ground water exchange pilot program in Montana', *Journal of Contemporary Water Research & Education*, no 148, pp33–43

5 Conflictive rationality and aquifer protection

Aquifer: a mysterious, magical and poorly defined area beneath the surface of the earth that either yields or withholds vast or lesser quantities of standing/flowing water, the quantity and/or quality of which is dependent on who is describing it or how much money may be at stake.

– R. Radden (2002)

The perception that conflicts or negotiations over groundwater are all about allocation and ownership is misinformed. Some of the most contentious battles over groundwater focus on the perceived threats to the quality of groundwater. One has to look no further than the media frenzy over the threat to local, regional and national water supplies associated with the hydrofracking debate. But community cohesion and civility can become fragmented and deepen the urban-rural divide when it comes to the issue of delineating protection areas for wellheads, springs, and recharge areas. This chapter describes a case study of a conflict and related negotiations regarding an aquifer protection area that has been underway for over 15 years. It describes how the politicization of science has diminished the notion of becoming an aquifer community. Application of the transdisciplinary imagination to the debate reveals a path forward relying on a successor effort by passing the aquifer protection baton to the next generation of aquifer users.

Collaborative rationality is a process where 'all affected interests jointly engage in face to face dialogue, bringing their various perspectives to the table to deliberate on the problems they face together. . . . [A]ll participants must also be fully informed and able to express their views and be listened to, whether they are powerful or not. Techniques must be used to mutually assure the legitimacy, comprehensibility, sincerity, and accuracy of what they say. Nothing can be off the table'. They have to seek consensus, or at least substantial agreement and 'find creative ways to satisfy all participants' (Innes and Booher, 2010, p6).

I have found that conflictive rationality is just the opposite. The affected interests independently engage in face-to-face dialogue using public meetings for deliberations. While the various interests are highly educated and informed about the scientific subject matter, neither listens to the other, and there are no techniques to assure legitimacy, sincerity or accuracy. Information is kept off the table.

Science is used as a strategic and tactical resource in ideological debate. Advocates of particular positions may 'spin', 'cherry-pick' or even misuse information to present their desired outcomes. Consensus or substantial agreement cannot be sought, as there is a lack of trust. Nobody is satisfied, and the 'politicization of science' ensues (Pielke, 2007).

The antagonism between urban residents who feel an affinity to the 'greater good' versus rural residents who value independence and wide open spaces and who 'just want to be left alone' is real. There is a common misconception from urbanites that rural residents are not sophisticated and are uneducated, conservative and anti-environment; conversely, rural residents often consider urbanites as snobs, morally corrupt and eco-centric liberals. These situations are generally identity-based conflicts where issues of identity, cultural values, worldview, dignity, recognition, safety, control, purpose and efficacy are more likely to lead to conflict between disputants than disputes over water or other tangible resources (Rothman, 1997). Rothman offers that while identity-based conflicts are destructive, once identified and processed with the right approach these disputes can evolve into creative outcomes with significant opportunities for 'dynamism and growth'.

The urban-rural divide is perhaps most noteworthy in the case of large water impoundment and diversion projects. Large dams displace rural residents. China's dam-building boom is reportedly displacing millions of rural residents. *Drowned Out* is a documentary film that describes the displacement of farmers in India as the Narmada Dam is constructed. The documentary *Waterbuster* describes the displacement of Native Americans during the dam-building boom on the Missouri River in the United States (see Appendix B). As we have seen in earlier chapters, much of the debate regarding the wellfield and pipeline project planned by the Southern Nevada Water Authority to supply water to Las Vegas focuses on the impact of groundwater pumping on rural residents located in north central Nevada and western Utah.

And, while the urban-rural divide on water issues on large water projects is a media darling, less well documented are the issues associated with groundwater. In 2012, the Michigan State Legislature passed Public Act 602, Aquifer Protection and Dispute Resolution, which was designed to protect the 'little' groundwater users from the high-capacity users. It does so by providing a formal process for resolving disputes between users of high-capacity wells (such as irrigation wells and municipal wells) and users of residential or other low-capacity wells who allege impacts from the large withdrawals. The conflict resolution field has evolved to the point where specialists such as the Rural Mediation Group (ruralmediation.com) in Dublin, Ohio, assist rural landowners with everything from grain contracts to dealing with energy companies on hydrofracking.

A new type of conflict associated with the urban-rural divide focuses on land use associated with protecting recharge areas located on private property or outside the jurisdiction of municipalities that use wells and springs as part of their drinking water supplies. These types of conflicts are becoming more commonplace with exurban development.

The following case study is unique in that it is a personal account from when I lived in Wyoming and worked as a groundwater professional with little training in process or conflict management. I became involved in a dispute over land use and groundwater protection at the junction of the urban-rural divide during the mid-1990s. After leaving the region, I followed the conflict in the media, while receiving advanced training and professional practice in water resources conflict resolution, transformation and negotiation. I later returned to the area of the conflict to conduct a stakeholder assessment as part of my research for this book. What may also be a unique aspect of this research is that I have volunteered, worked and interacted socially with nearly all of the stakeholders since the 1990s. The paradox is that participants on both sides of the urban-rural divide are highly educated and experienced within the fields of geology and groundwater, that conservatives and liberals reside on both sides of the 'buck-and-pole' fence and that there is a mutual respect and desire for open spaces and protecting the environment.

Personal accounts of conflict or collaborative experiences by water 'pracademics' are not common, but there are a few recent examples by Moore (2013), Gyawali (2013), Daniels and Walker (2012), Margerum (2011) and Innes and Booher (2010). What one quickly discovers is there is not only a divide between urban and rural factions, but also a rift zone between what academics, civic movement activists and experts studying water think and where politicians and their everyday politics drive decisions on water. I will lay out below what I experienced from within the 'rodeo ring' of water conflict transformation and negotiations with the blood sport of aquifer protection, and then reflect on some lifelong lessons.

Overview of motivation for groundwater protection

Federal water development projects were undergoing a boom during the 1950s, fitting with what Freeze (2000) describes as the 'Age of Carelessness' and the 'Throwaway Society'. Changes in the perception of natural resources and their use that occurred during the 'Age of Awakening' in the 1960s led to the establishment of the Environmental Protection Agency (EPA) (Freeze, 2000). The importance of groundwater as a national resource prompted the formation of the National Ground Water Association, which began publishing the journal *Groundwater* in 1963.

Changing environmental perspectives and regulations during the 1970s prompted the 'Age of Awareness and Action' as described by Freeze (2000). Freeze (2000) characterizes this period as one in which regulations governing clean water, safe drinking water and resource conservation were established. For example, the Safe Drinking Water Act (SDWA) of 1974 was enacted, which set standards for the quality of drinking water.

The 1980s were described as the 'Age of Disillusion' by Freeze (2000), as the 'Superfund' legislation became law. The SDWA, as amended in 1986, required public water supply systems to determine wellhead protection areas (WHPAs) for wells and springs used as drinking water supplies.

Reflecting on professional experience spanning over 20 years after publishing the landmark textbook *Groundwater* (Freeze and Cherry, 1979), Freeze (2000) suggested that some steps in the right direction of future groundwater policy included (1) developing policy emphasizing prevention of future contamination, (2) siting waste management facilities properly and (3) stressing aquifer protection and watershed control rather than site cleanup.

Case study and hydrogeologic overview

The following case study is located in the city of Laramie, Wyoming. Laramie is a good selection for a case study on conflict over groundwater and aquifer protection because it hosts the only four-year university in the state of Wyoming, which is famous for its engineering, geology and geophysics academic programs. Laramie is also the home base for many environmental consulting firms. And Laramie typifies a community with an urban base that relies on both surface water and groundwater supplies, with surface water piped from distant sources west of the city; wells and springs are located close to the city, yet the recharge area for the groundwater sources are beyond the jurisdictional limits of the city. Exurban subdivisions that were built in the late 1960s and continue to develop are located close to Laramie and rely on individual wells tapping the same aquifer as the city for water supplies. Rural residents rely on on-site wastewater systems that consist of buried septic tanks and a soil absorption system. One of the main interstate highways in the United States, Interstate 80, is located in Telephone Canyon, which is incised from the bottom to the top of the Casper Aquifer, a regionally extensive aquifer system composed of layers of limestone and sandstone about 213 m thick. The limestones are locally mined for manufacturing cement.

Laramie was sited by Union Pacific Railroad surveyors to take advantage of large springs that discharge from the Casper Aquifer. These are localized along prominent fracture zones (Huntoon and Lundy, 1979). Lesser springs discharge from all the permeable strata exposed low along the flanks of the Laramie Range, including small amounts of water from the lower permeability redbed sequences within the Satanka redbeds that overlie and confine the Casper Aquifer. The origin of the water is locally derived recharge along the flank of the Laramie Range that circulates westerly toward the Laramie River.

The city of Laramie provides water to roughly 95% of the 37,000 people living in Albany County. The Casper Aquifer supplies all of the water to approximately 400 rural residences located on the flank of the Laramie Range in Albany County and approximately 60% of the municipal water supply for the city of Laramie (Wittman, 2008). Laramie acknowledged the importance of protecting the groundwater supplies situated on the eastern side of the Laramie Basin by obtaining a grant from the EPA in 1993 to develop a wellhead protection plan (WHPP; see Table 5.1). Laramie selected a local consulting engineering firm to develop the WHPAs. A WHPP draft ordinance developed by Laramie was also a condition of the EPA grant.

Table 5.1 Timeline of aquifer protection activities

Year	Description of aquifer protection activity
1993	• City of Laramie receives grant from EPA for Wellhead Protection Plan (WHPP) because 60% of drinking water supply is from groundwater stored in Casper Aquifer. • Local consultant delineates WHPP for municipal wellfields and springs.
1996	• Draft Wellhead Protection Ordinance developed by city. • Citizen comments suggest peer-review of WHPP.
1997	• Laramie City Council and Albany County Commissioners task the Environmental Advisory Committee (EAC) to develop Aquifer Protection Plan using community volunteers.
1998	• EAC appoints volunteer committees. • Technical Advisory Committee, comprising engineers, geologists, hydrogeologists and large acreage landowner, completes Casper Aquifer Protection Area (CAPA) delineation.
2000	• Albany County Commissioners and Laramie City Council sign Joint Resolution supporting EAC Delineation Report and Map and Contingency Plan and request EAC develop an Aquifer Area Management Plan. • City of Laramie designated a Groundwater Guardian community by the Groundwater Foundation.
2002	• Wyoming Department of Environmental Quality (WDEQ) reviews Casper Aquifer Protection Plan (CAPP). • City of Laramie adopts Aquifer Protection Overlay (APO) Zone ordinance and Albany County adopts APO Zone resolution. • City of Laramie hires water outreach coordinator.
2006	• WDEQ completes review of CAPP. • University of Wyoming Jacoby Golf Course proposes expansion and adjacent subdivision development. Citizen petition opposes expansion on basis of proximity to water sources, pesticide/herbicide/fertilizer use and CAPP enforcement. • Citizens for Open Space and Outdoor Recreation file petitions opposing development. • EAC recommends denying annexation and permit for development based on nitrate loading study and City Draft Comprehensive Plan expressing desire to preserve open space and protect ridgelines from development. • Building moratorium imposed until a modified Aquifer Protection Plan and associated ordinance requiring environmental reports for new development in APO Zone are adopted. • City council authorizes update of CAPP.
2007	• City Groundwater Guardian designation lapses. • Out of state consultant hired by city of Laramie to update CAPP. • Lawyers for large landowners correspond with city council about legal 'takings', lack of new scientific data, goals of citizen petitions, and extraterritorial jurisdiction. • City council authorizes submitting revised CAPP to WDEQ. • Building moratorium extended to one year. • WDEQ approves revised CAPP.

Table 5.1 (Continued)

Year	Description of aquifer protection activity
2008	• City's consultant submits updated CAPP, overlay zone and revised ordinance.
	• City's consultant submits estimate of Casper Aquifer recharge study. Estimate average annual recharge of approximately 1 inch for the last 26 years and a long-term trend of decreasing recharge.
	• Citizens contact WBPG regarding unlicensed practice by city's consultant.
	• City council adopts updated CAPP, overlay zone, and revised ordinance. Western boundary of protection area changed. CAPP not certified by state licensed geologist or engineer.
	• Three local state licensed geologists send letters of support to city's consultant after CAPP submitted to city.
	• 'Cease and desist' letter sent from WBPG to city's consultant.
2009	• Formal complaint regarding unlicensed practice filed with WBPG by private citizen.
	• Citizens for Clean Water actively lobby WBPG in support of city's consultant.
	• City samples domestic wells in CAPP for nitrate-nitrogen and presents report.
	• Nitrate-nitrogen data reported for domestic wells sampled by private citizen.
	• WBPG refers results of investigation to Albany County prosecuting attorney.
	• County Planning and Zoning Commission begins updating a version of CAPP without boundary changes adopted by city.
2010	• City resamples some wells in CAPP. Memo released to city council with generalized results of well resampling.
	• City retains a new consulting team to assess risk to city's wells and springs from residential septic tanks. Risk assessed as 'low'.
	• Citizens living in Casper Aquifer Protection Area form CAP-Network and join Groundwater Guardian.
	• City reestablishes Groundwater Guardian affiliation.
	• Science panel meets at University of Wyoming to discuss history, geography and science of CAPP program.
2011	• CAP Network begins own well water sampling program.
	• City's Groundwater Guardian affiliation lapses.
	• City hires new water resources specialist.
	• City and county propose land swap of 50,000 acres at Y Cross ranch with University of Wyoming (UW) Board of Trustees for 10,000 acres of land in and around CAPP.
	• UW Foundation indicates swap is incompatible with agreement with Y Cross ranch and joint owner, Colorado State University.
	• Junior high student completes project on background levels of nitrates in the Casper Aquifer using Klein Spring at 1.8 ppm (mg/L) nitrate-nitrogen.
	• Interstate 80 mitigation plan for spills presented to city, county and WDOT.
	• County commission approves an Aquifer Protection Plan different than the city's with no change in western boundary of protection area.

(Continued)

Table 5.1 (Continued)

Year	Description of aquifer protection activity
2012	• Legislator proposed state of Wyoming purchase 11,000 acres of land within CAPP for $15 million.
	• Wyoming Water Development Office director pans land purchase concept in press.
	• County Planning and Zoning Commission proposes changes to boundary of CAPP.
	• Local geologists write petition signed by property owners countering County Planning and Zoning Commission recommendations; file petition with county commissioners.
	• Citizens for Clean Water file petition with county commission and city council regarding flaws in county resolution.
	• Junior high student completes continuation project exploring native nitrogen-fixing plants as a source of natural background levels of nitrates in the Casper Aquifer. Confirms soils near naturally growing nitrogen-fixing plants such as Mountain Mahogany and Cryptogamic crust produce elevated concentrations of nitrates as high as 200 ppm in water leached from soils.
	• Water resources specialist resigns.
	• Interstate 80 Aquifer Protection Monitoring Well Plan and Detention Pond Design team selected.
	• County commission approves regulations that differ from city's, specifically on maintaining western boundary defined in 2002.
2013	• House Bill 85 passed repealing state law allowing city health ordinances within 5 miles of city boundary.
	• County commissioner suggests bill was in response to how Casper Aquifer was regulated.
	• City hires new water resources specialist.
	• Consulting team final report released indicating no significant risk of nitrate contamination from private septic systems to city water supplies. Predict more than a century would pass before the concentrations reached the 5 mg/L level at city water supplies.

Wellhead protection areas

A wellhead or source water protection area as defined by the U.S. and Canadian governments is the surface and subsurface area around a well, spring or tunnel through which contaminants are reasonably likely to move toward and contaminate the drinking water source. The dimensions of protection zones are variable, based on individual state and provincial needs, but typically approach the following spatial and temporal criteria:

• Zone One – the area within a 100-foot radius around the wellhead, spring or tunnel collection area

- Zone Two – the 180-day to 5-year time of travel (TOT), the boundary of the aquifer(s) that supplies water to the source or the groundwater divide, whichever is closer
- Zone Three – the area within a 3-year to 20-year TOT of the source, the boundary of the aquifer(s) that supplies water to the source or the groundwater divide, whichever is closer

Many protection zones are transient capture areas routinely determined by public domain computer software developed for ideal aquifers that have homogenous and isotropic hydraulic properties. They represent areas where different types of land use are not permitted. For example, as during the colonial period of the U.S. settlement, potential contaminant sources are prohibited within a fixed radius of a well or Zone One. The other zone protection areas are designated to prohibit other land uses such as 'the necessities of nature' or locating gas stations (Witten et al., 1995).

Hydrogeology and influence on wellhead protection areas

The areal distribution of aquifers is a function of the geology of a region. In areas underlain by sand and gravel, the hydrologic boundaries of the sand and gravel aquifers are typically the vertical and lateral extent of the porous materials. In areas underlain by bedrock such as sandstone and limestone, the areal extent of the aquifer is not only defined by how the rocks are folded and faulted, but also by how much of the rocks are saturated. For example, bedrock aquifers located along a mountain range may be only partially saturated where the rocks outcrop and recharge occurs, whereas the areas where the rocks disappear beneath the land surface due to tilting may be fully saturated with water. The hydrologic boundaries of protection areas in bedrock aquifers are complicated by the degree of saturation, the volume of storage space within the rocks and the juxtaposition of permeable rocks against impermeable rocks in areas of faulting and areas of enhanced permeability, where hard rocks have undergone brittle deformation by the fracturing associated with folding and different types of faulting, or areas of lower permeability, where rocks have undergone ductile, or plastic, deformation. The assessment of increased or decreased permeability associated with structural deformation of rocks follows some very generalized rules, and detailed assessment typically is undertaken on a case-by-case basis through drilling and well-testing programs.

Groundwater moves from areas of high hydraulic head toward areas of low hydraulic head; in the case of the Casper Aquifer this is from the highlands east of Laramie westerly toward the Laramie River. Under natural conditions, water in the aquifer flows from the recharge area toward the discharge area. Springs represent a discharge area, or areas of low hydraulic head; but for the spring to continue to flow, groundwater flows from storage within the aquifer toward the spring by gravity. Wells drilled into aquifers capture water stored in the aquifer by lowering the hydraulic head in the vicinity of the well during pumping to create artificial discharge areas. The zone of influence or pressure change surrounding a well is called the radius of influence. Water discharged from a well is derived by capturing water stored in the aquifer. Groundwater flowing toward the well during pumping is derived from storage in the aquifer and is 'captured'; the 'capture area' can be estimated and mapped.

The size of both the radius of influence and the capture area is not only a function of the hydraulic properties of the aquifer, but also of the pumping rate and the duration of pumping. The geographic significance of both the radius of influence and the capture area is that these can become management areas, requiring policies directing how many wells can be drilled into an aquifer in order to efficiently exploit the water stored in the aquifer, protect from overdrafting or protect the aquifer from contamination due to land use within the capture area (Livingstone et al., 1996).

Challenges to the wellhead protection plan

With this background, Laramie's consulting engineer completed the delineation of the WHPAs for the municipal wells and springs in 1993, using the method of hydrogeologic mapping and TOT contours defined by major faults, under the assumption that faulting and folding locally enhanced groundwater circulation within the limestone and sandstone comprising the Casper Aquifer. Two years later, a large-animal veterinary clinic purchased property located on the Satanka redbeds, which were considered a low-permeability shale that confined the underlying Casper Aquifer, to construct a building serviced by a water well supplied by the Casper Aquifer. A conventional septic-tank wastewater-treatment system was anticipated to be installed in the redbeds. While the location of the clinic, water well and on-site wastewater system were located outside of the boundaries of the delineated WHPAs, Laramie's consulting engineer challenged the location of the facility on the basis of the proposed septic system's proximity to a buried fault. Laramie's concern regarding conventional septic systems focused on the premise these wastewater systems are a source of nitrate. The veterinary clinic retained the engineering firm that employed me to collect hydrogeologic data to allay the concerns of Laramie's consultant. The dispute evolved to become a 'wicked problem', as nearly every source for conflict was triggered (see Figure 5.1). Even though the situation remained at an impasse between the dueling experts, as depicted in Figure 5.2, the clinic was constructed after negotiations for the regular monitoring of the site water well for nitrates.

A public meeting was held in 1996 to introduce the Wellhead Protection Ordinance and delineated areas. Citizens voiced numerous concerns at that time, based upon (1) the prescriptive nature of the ordinance, (2) the dependence of the 1993 WHPAs upon the location of identified faults and (3) the exclusion of limestone quarries from the WHPAs (Wittman, 2008). One of the largest property owners in the area was also concerned because the delineated areas fell on private property outside the jurisdiction of the city of Laramie. And, given the situation with the veterinary clinic and the dispute over the existence and

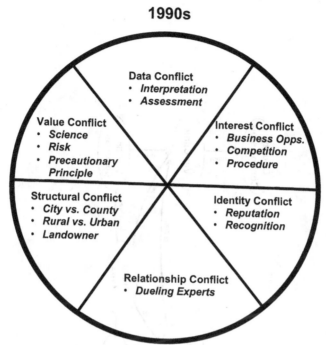

1990s

Figure 5.1 Wellhead protection Circle of Conflict

hydrologic role of a 'buried' fault that fell outside of the boundaries of the delineated WHPAs, there were concerns by county residents regarding property rights.

In 1997, the Laramie City Council and Albany County Commissioners appointed me to the Environmental Advisory Committee (EAC). The EAC was tasked with developing an aquifer protection program for the Casper Aquifer, as it was clear that many residents outside the urban boundaries of Laramie relied on the aquifer for their sole source of drinking water supplies. A couple of important changes to the groundwater protection programs across the United States and in Wyoming had also occurred about this time: (1) the SDWA was amended in 1996 to acknowledged the interaction of surface water and groundwater, and it required public water systems to refine their wellhead protection programs to source water assessment or aquifer protection programs, and (2) the Wyoming Department of Environmental Quality (WDEQ) received a grant from the EPA to develop a statewide Wellhead Protection Program guidance document. The engineering firm where I was employed was selected to prepare this document under contract to EPA. Given our experience elsewhere in the western United States, we developed the statewide plan to align with the recent amendments to the SDWA, with a focus on a more holistic approach to protecting groundwater through an aquifer protection program.

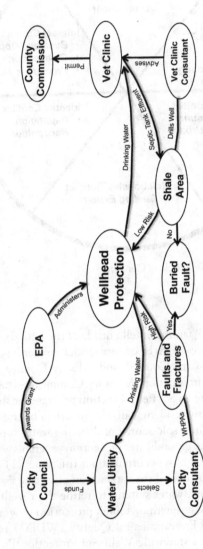

Figure 5.2 Wellhead protection situation map

The western boundary problem

The first step that the EAC took in 1998 was to be more inclusive of urban and rural residents' participation in the aquifer protection program. Advertisements were placed in the local newspaper soliciting letters of interests from residents to work not only on the 'technical review committee' tasked with delineating the aquifer protection areas, but also on subcommittees that included the public education, contaminant source inventories, contingency planning and aquifer area management. Over 30 volunteers applied for the various committees.

The technical review committee was composed of volunteer engineers and geologists from the various engineering firms located in Laramie, as well as professors in geology and geotechnical engineering from the University of Wyoming. Even though they admittedly lacked the technical skills to serve on the technical review committee, the largest property owners in the area also participated, given their extensive local knowledge of the recharge area. It was also important to the community that the landowners fully understand the delineation of boundaries by the technical review committee. The city's consultant, who was responsible for delineating the WHPAs, and I both served on the technical review committee. All meetings of the technical review committee were informal and were held at public locations such as the county library.

The boundaries of the Casper Aquifer Protection Area were initially delineated on the basis of agreement as to what parts of the aquifer served as the principal water bearing zones, followed by agreement on lesser priority zones based on lower productivity and depths of the municipal wells, as well as the elevation of the major springs developed in the county. Consensus was reached that the eastern boundary included the watershed divide of the Laramie Mountains, the northern boundary was located north of a large fold known as Spur Anticline, the southern boundary was located south of the Simpson Springs Anticline and the western boundary was *calculated* from a dip formula, where it was estimated that approximately 23 m of Satanka redbeds would overlie the Casper Aquifer. The definition of the western boundary was the most problematic, as there was little agreement on the hydrologic role of the redbeds. The city's consultant was adamant that the hydraulic properties of the Satanka redbeds were not homogeneous along the stratigraphic contact between the Casper Aquifer and the overlying redbeds, so a safety factor was needed to ensure aquifer protection. The calculated approach was developed by the technical review committee as a compromise to those with opinions contrary to this position and in the spirit of cooperation and conciliation.

The agreed-upon delineated aquifer protection area encompassed approximately 186 km^2; the report was signed by the technical review committee members on July 25, 1999. The delineation report was presented at a joint work session of the Albany County Commissioners and the Laramie City Council. On January 4, 2000, both governing bodies approved the delineation through a joint resolution. The initial Casper Aquifer Protection Plan (CAPP) was completed and submitted to the WDEQ in 2002 (Wittman, 2008). In the interim, one member of the technical review committee and I departed from the area. Volunteer successor efforts through the EAC were replaced with a water coordinator hired by the city to lead the implementation efforts during the course of the WDEQ review, revision and resubmittal, which was completed in 2006 (see Table 5.1).

Even though I had left the area in the late 1990s, the progress of the Casper Aquifer protection program intrigued me, as it was my first experience with the nexus of science and policy. It was relatively easy to track the issue since it was regular news in the local newspaper that could be read online. Given the importance of the aquifer to Laramie's drinking water supply, it was also a regular topic on the agendas of both the city council and county commissioners, whose agendas, meeting minutes and videos were posted on their respective websites. The local newspaper also permitted anonymous comments to articles and letters to the editor that provided valuable feedback on the community's opinions on the aquifer protection program. It was not unusual for an article related to this topic to receive 20 to 40 comments.

On the basis of media reports and meeting minutes, the conflict over the western boundary of the Casper Aquifer Protection Area resurfaced in 2006, when the University of Wyoming and a land developer proposed an eastward expansion of the university's golf course and associated subdivision. Citizens who favored open spaces and increased access to recreational opportunities on the recharge area of the Casper Aquifer filed petitions against the development based on concerns over the use of chemicals and fertilizer in the vicinity of the Casper Aquifer Protection Area. Citizens for Open Space and Outdoor Recreation, the first of a few nongovernmental organizations (NGOs) dedicated to land use and property rights on the Casper Aquifer, was formed. It conducted the first of many petition drives to convince the city council that additional environmental investigations were needed before permitting any development along the western boundary of the aquifer protection area. The EAC expressed concern over ridgelines, viewsheds and nitrate loading from the golf course in the aquifer protection area. A building moratorium was imposed by the city council until the aquifer protection plan was modified, and a new ordinance was passed that required environmental reports for proposed development in the Aquifer Protection Overlay (APO) zone. The city council also authorized plans to update the CAPP using a consultant to review and revise the boundaries of the aquifer protection area designated by the local volunteers. All of these efforts to change the aquifer protection plan occurred before the WDEQ approved the first aquifer protection plan in the State of Wyoming.

And while the city submitted the revised aquifer protection plan to the WDEQ, the city also retained the services of an out-of-state consultant to update the CAPP. About the same time, the previously imposed building moratorium was extended, much to the chagrin of the subdivision developers. The first inkling of potential litigation over property rights surfaced when attorneys for property owners within the delineated aquifer protection area corresponded with the city council about 'takings' and extraterritorial jurisdiction, among other related concerns.

Even with the 2007 WDEQ approval of the revised CAPP, the city's consultant submitted yet another revised CAPP, a revised ordinance and an aquifer recharge study to the city in 2008. The western boundary of the CAPP was extended by the consultant to the west of the boundary delineated by the EAC technical review committee on the basis of easier implementation, in order to provide an additional buffer beyond the safety factor incorporated by the EAC technical review committee, and on the basis of new interpretations of geologic data. Rather than relying on the calculated estimate of the buffer of protective lower-permeability shale, the updated boundary was based on legal boundaries of section, quarter-section and quarter-quarter section lines. However, the consultant also indicated that 'Significant technical changes to the delineation of the Casper Aquifer Protection Area boundaries will be reviewed and approved by three Wyoming licensed professional engineers or geologists' (Wittman, 2008). The city's consultant did not have a Wyoming-licensed professional engineer or geologist on its staff. An inquiry about the unlicensed practice of geology before the public was made by a local geologist with the Wyoming Board of Professional Geologists (WBPG). While the complaint was under review by the WBPG, three geologists who were listed in the acknowledgements of the consultant's report wrote letters of support to the consultant after the CAPP was submitted to and approved by the city. Later in 2008, the WBPG sent a cease-and-desist letter to the city's consultant. A former University of Wyoming geology professor and EAC technical review committee member reflected on 30 years of scientific work, as well as living on the Casper Aquifer, and opined that 'the issue is about a community working together to protect itself and ensure its survival' (Stauffer, 2008). In a 2008 letter to the Albany County Planning and Zoning Commission, a hydrogeologist that was a past student of the same professor, and who also lives on the Casper Aquifer, wrote in response to Stauffer's article, 'But I have to ask, if one opposes the grassroots "protect our groundwater" organizations, does that mean that he or she is against clean water?' (Richter, 2008). The urban-rural rift zone was just beginning to open and the Circle of Conflict was starting to look more like the loaded chambers of a six-shooter pistol (see Figure 5.3).

In 2009, a rural resident filed a formal complaint regarding the unlicensed practice of geology with the WBPG. A new NGO, Citizens for Clean Water, actively lobbied against the complaint lodged against the city's consultant. During this same year, the city sampled rural residential wells for nitrate-nitrogen through volunteerism and anonymity. Samples of some wells were also independently collected by a rural resident. The city's sampling results for many rural residential wells in the 1960s-era subdivision close to the city boundaries indicated nitrate-nitrogen at concentrations greater than the 10 mg/l drinking water standard for public drinking water systems. The rural residents challenged the results, however, based on sample and analysis integrity. It is important to recall that many of the residents in the rural subdivisions are trained in geology and engineering: some work for local consulting firms and some work as faculty and staff at the University of Wyoming. Later in that year, the WBPG completed its investigation of unlicensed practice of geology before the public and forwarded the result to the county attorney. The County Planning and Zoning Commission also commenced updating a version of the CAPP that did not acknowledge the changes to the western boundary adopted by the city. Clearly, the trust between the city and county had been broken, and the urban-rural divide had deepened.

Figure 5.3 Aquifer protection Circle of Conflict

Acknowledging that a single round of well sampling requires verification of the analytical results, the city resampled some of the rural residential wells in 2010, although many of the owners of the previously sampled wells did not participate in the resampling program, due to waning trust. The city retained a new consulting team to assess the risk the rural subdivisions posed to the municipal wells and springs based on the existing water quality data for the municipal wells and springs and the sampled rural residential wells, and by integrating the hydrogeology of the Casper Aquifer, as it was known at the time.

Another new NGO formed: the Casper Aquifer Protection, or CAP, Network. It was initiated by some of the rural residents living on the Casper Aquifer and who relied on individual domestic wells for their water supply and onsite wastewater treatment systems. Both CAP Network and the city joined or reestablished their affiliation, respectively, with Groundwater Guardian, a national program devoted to protecting groundwater quality.

The science panel situation

As the level of mistrust increased between urban and rural residents, the time dedicated to public comment at city council and county commission meetings was increasingly directed toward the debate over the Casper Aquifer protection program. Much of the debate could be described as a 'dueling experts' or what I have referred to as 'guerrilla well-fare', with each party providing conflicting conceptual models and interpretations of data (Jarvis, 2010). In 2010, the Laramie/ Albany County Groundwater Guardian Team, in concert with the University of Wyoming, sponsored a science panel where two hydrogeologists who resided in Laramie were to debate two hydrogeologists who resided in the rural subdivision located on the outskirts of Laramie. Both 'teams' were identified as professionals who regularly spoke on Casper Aquifer issues at city and county forums. The science panel was advertised as meeting in a classroom located at the university; a mediator from one of the environmental research institutes at the university would moderate the debate.

As the planning for the science panel progressed, a hydrogeologist residing in the county learned that the mediator met in private with the 'team' residing in Laramie. The rural hydrogeologist considered this a breach of trust and considered not participating in the science panel, as he felt that a public ambush might be one of the outcomes. This same hydrogeologist contacted me and requested my participation in the science panel, given my past history with the EAC technical review committee that had prepared the initial delineation of the Casper Aquifer Protection Area. I consented to do so, as (1) I knew all of the parties on the science panel through academic, professional and social relationships and (2) I would focus my contributions on the history of the aquifer protection program and ways to move the discussions forward in terms of resolving the technical differences, rather than revisiting the past.

I was surprised by the recommendation of the university and the Laramie hydrogeologists to hold a science panel as a means to resolve the scientific issues surrounding the protection of the Casper Aquifer. Pielke (2007, p148) suggests that it 'is naïve to think that science advisory panels deal purely with science. Such panels are convened to provide guidance on policy, or on scientific information that is directly relevant to policy.' The science panel did not resolve anything related to policy, however, as policy discussions were outside the ground rules that the mediator developed at the beginning of the panel discussion. The outcome was just the opposite. The event added a new layer of mistrust as the university mediator lost the trust of one of the key participants before the panel began – the hydrogeologist residing in the rural subdivision, who was serving as the pro bono scientist working with the newly formed CAP Network.

But while the science panel did not strictly deal with resolving some of the disputed science associated with the Casper Aquifer, it did serve as an opportunity for collaborative learning. As reported by Haderlie (2010), the panel discussed different conceptual models to explain the elevated nitrates found in private wells located within and near the western boundary of the Casper Aquifer Protection Area. Clearly, the dispute was over what constituted a 'problem'. For example, the city's water coordinator indicated that 'nitrates can develop from natural sources, but elevated levels may indicate anthropogenic sources' (Haderlie, 2010). The hydrogeologist working with CAP Network indicated that while '[i]t has been stated that septic systems are the cause for elevated nitrates . . . I am looking at it from a different direction, that the wells are actually the means by which nitrates are getting into our aquifer' (Haderlie, 2010). One of the hydrogeologists living in Laramie offered, 'Certainly wells are part of the problem, whatever your nitrate source is' (Haderlie, 2010). The other hydrogeologist, who worked pro bono for one of the other NGOs and lives in Laramie, opined, 'The fact that we have high nitrates where we have a high concentration of septic [systems] makes a very compelling case, just by simple logic of association, that the septics are the source of the nitrates' (Haderlie, 2010). On the basis of nitrate issues elsewhere in Wyoming, as well as in Oregon, I offered yet another explanation: 'fertilizer [causing elevated nitrate levels] that was coming not only from farming but also from lawns. . . . I think it is somewhat of a leap to just concentrate all the emphasis on the septic tanks' (Haderlie, 2010).

Looking back on the science panel held in the public format, it was clear that 'politicization of science' had occurred, perhaps without the direct knowledge of the hydrogeologists participating in the science panel. Pielke (2007, p63) argues that 'a great danger for both science and politics occurs when members of the scientific community itself participate in the "politicization of science" through the media.' In the media 'contest', it is difficult to tell whether science or politics are on stage, and the power of science and credibility of the democratic process is obscured by the process. 'Loss of the power of science matters only if science, or more accurately the information provided through science, has in some cases a unique role to play in the policy process. For if information does not matter, then distinguishing science and politics would be of little concern; we could all then simply invent "facts" as convenient' (Pielke, 2007, p63).

The new identity

The science panel situation did little to persuade either coalition to change its positions regarding CAPP. The CAP Network initiated its own residential well sampling program for nitrates and did not share its data with the city. A middle school student residing in the rural subdivisions built in the Casper Aquifer Protection Area completed an investigation of background concentrations of nitrate-nitrogen, and won an award for the project at a science fair; however, the city was apparently not interested in learning more about the research or this local knowledge.

A 2011 proposal to swap private property located in and near the Casper Aquifer Protection Area with property elsewhere in Wyoming that was jointly owned and operated by the University of Wyoming and Colorado State University provided the first glimmer of renewed interest in the city and county working together since the late 1990s. The city and county co-proposed a swap of 11,000 acres of private land underlain by the Casper Aquifer for

50,000 acres of land located farther east of the Casper Aquifer Protection Area, which had been donated to the universities. In the eyes of the competing interests on aquifer protection, this appeared to be a win-win proposal for all parties; however, the University of Wyoming quashed the proposal as incompatible with the agreement by the donor and the joint operator, Colorado State University.

While the threat of a potential spill of chemicals along an interstate highway has been recognized since the completion of the 2002 wellhead protection plan, it was not until 2011 that a formal mitigation plan was completed and presented to the city and county. The plan recommended installing monitoring wells and the construction of detention ponds to capture any spilled chemicals or fuels along a portion of the U.S. Interstate Highway System; this section is one of the highest points of the major road system and is known for severe winter weather and poor driving conditions. Although both the city and county apparently agreed on the need for additional monitoring and spill mitigation, the county commission approved a Casper Aquifer Protection Plan that differed from the city's approved plan. Notably, the county's version did not change the location of the western boundary of the protection area.

The city and county were not dissuaded from pursuing the concept of securing the Casper Aquifer Protection Area, despite the setback with the University of Wyoming. In 2012, a local legislator introduced a bill to have the state of Wyoming purchase 11,000 acres of land within the Casper Aquifer Protection Area for $15 million, with the idea that the land purchase would not only serve to protect the city and county groundwater, but also become the newest addition to the state parks system. This was an important change in the conflict and one that was recognized as transformative when the sponsoring state legislator indicate he believed that the resolution of the aquifer protection issue was necessary to allow future development of the community (*Billings Gazette*, 2012).

The transformation of the conflict by the notion of a land purchase moved the boundary war from the views of entrenched interests to seeing that the basket of benefits that might be developed for the benefit of all. The collaborative learning emphasis is on capacity and competency building, primarily of institutions such as the city or the county. But van Vugt (2009) indicates that identity works toward action by connecting groups of competitors to move toward action. He suggests that it is important to create superordinate identities, such as regions, by thinking of ways to 'blur group boundaries'. This can be done by referencing a concept such as, 'we all drink from the Casper Aquifer', implying that 'we are all in this together'.

Although citizens in both the city and county approved of the concept and lobbied the state legislature for the land purchase, the director of the principal water development agency for the state panned the concept because 'land outside the proposed acquisition area could still affect the quality of the city's well water . . . [and] residential development itself may not be a concern. . . . A lot of water comes through that recharge area' (*Billings Gazette*, 2012).

Boundary wars return

The land purchase did not get funded. So the battle over boundaries resumed, with the County Planning and Zoning Commission proposing changes to the western boundary of the aquifer protection area. Articles and letters to the editor in the local news were frequent; they often referred to the impacts of the changes in the western boundary on longtime businesses located near the boundary, and they were often supplemented by tens of comments both for and against the proposed changes (Newman, 2012). A group of geologists, who resided in the rural subdivisions located in the aquifer protection area, including the geologist associated with CAP Network, submitted a petition to the county commissioners countering the recommended changes. Citizens for Clean Water countered with a petition describing flaws in a county resolution that did not include the change in the western boundary. When the smoke cleared on the debate, the county commission approved regulations that again differed from the city's: namely, that the western boundary of the aquifer protection area would remain the same as originally delineated in 2002.

What was lost in the boundary debate was an important study by the same middle school student who studied the background concentrations of naturally occurring nitrate, the principal contaminant of concern throughout all of the boundary wars. This time he determined that soils found near nitrogen-fixing plants produced elevated concentrations of nitrates, on the order of 20 times the contaminant level of 10 mg/l. Once again, local knowledge fell victim to the politicization of science.

While the protection of the Casper Aquifer was apparently seen as politically neutral and morally justified by the urban environmental community, the rural residents living on the Casper Aquifer considered the efforts to be little more than an attempt to wrest ownership or 'take' their property rights from them. In 2013, a local state legislator introduced House Bill 85, which would repeal a state law permitting a city to exercise extraterritorial jurisdiction, allowing mayors to enforce city 'health, or quarantine ordinance and regulation thereof', within 8 km of city borders. According to one county commissioner, the bill stemmed from differences in how the Casper Aquifer was regulated: 'The city and council worked together, but we had a couple of city councilmen who were politicking' (Hancock, 2013). Later in the year, the city's consultant released a final version of the draft report completed in 2010, which indicated that the private septic systems located in the rural subdivisions built near the western boundary of the Casper Aquifer Protection Area did not pose a significant risk of nitrate contamination to the city's drinking water supply wells at this time. WWC Engineering (2013, p5) went on to state that the 'long-term monitoring data from the well-fields from as early as 1973 through May 2012 indicate that the concentrations of nitrate as nitrogen (NO_3 as N) and nitrate plus nitrite as nitrogen (NO_3+NO_2 as N) average from 1.37 to 1.91 mg/L. . . . The concentrations at the Turner and Soldier wellfields exhibit trends of slight increases over time, based on statistically significant correlations (at the 95% confidence level) between sample date and nitrogen concentrations. . . . If concentrations increased at those rates, more than a century would pass before the concentrations reached the 5 mg/L level at which increased monitoring is required'. Likewise, the concentrations of nitrate as nitrogen at the city's Pope and Spur wellfields also did not exhibit increasing trends.

Stakeholder assessment

Given the lack of trust between all parties since the first delineation of the Casper Aquifer Protection Area in 2002, I revisited the area in 2012 to investigate the possibility of a consensus 'rebuilding' process. Islam and Susskind (2013) describe a stakeholder assessment as an early step in the consensus-building process that is used to identify appropriate stakeholder groups and their representatives. During the course of the 2012 Water Diplomacy Workshop held at Tufts University, Lawrence Susskind indicated the stakeholder assessment could be conducted at any time during the course of conflict, so I elected to complete the stakeholder assessment, given that I knew all of the parties involved in the Casper Aquifer situation and that I had not lived in the area since 1998.

My goal was to conduct semistructured interviews of stakeholders from the various factions who were also people that I knew both personally and professionally when I lived in the area. Many of the interviewees were also geologists and engineers, even though they sat on different sides of the table on the issues. I informed each person that they would not be identified beyond their affiliation with a particular public or private entity. The interview was open ended, with me asking the same generalized questions of each person. I interviewed individuals from the city council, county commission, University of Wyoming, Citizens for Clean Water and CAP Network, as well as a volunteer with the County Planning and Zoning Commission.

What are your concerns?

The representative from the city council was concerned about the integrity of old wells and old septic systems that had undergone no inspection when the rural areas located on the Casper Aquifer were developed. According to the representative, common sense suggested that the observed concentrations of nitrates in some private wells were associated with septic tanks. The representative was also concerned that the housing market in the rural area built on the Casper Aquifer, as well as some businesses located near the aquifer were being negatively impacted by anecdotal reports of contamination issues, perceived threats of aquifer contamination and news articles. There was also concern that aquifer protection was being used as leverage for open space.

The representative from the county commission was concerned because the city openly criticized the county for decisions on development, but then adopted development plans that were contrary to the position critical of the county. Given the influx of new residents, there was also a lot of learning about the aquifer protection issues by the new residents that created conflict because of contradictory information shared by word of mouth and by the media. The aquifer is incised from the bottom to the top by the interstate highway – a major transportation corridor for hazardous materials. The issues of the boundaries associated with the protection area remain problematic, due in part to the general lack of reliable borehole data and the fact that state government could certainly help with the situation by logging the boreholes.

A citizen volunteer on one of the county advisory boards was concerned about continued development overlying one of the best groundwater sources in the state. The volunteer asserted that no one within the city or county was making decisions on the purity of the aquifer, and that there appeared to be more emphasis placed on deeper drilling than on protecting the state's groundwater resources. The selective use of information by residents living in the rural area clouded the rational discourse on which direction the city and county should move in order to protect the aquifer.

Residents of the city who actively participated in Citizens for Clean Water were most concerned about the contamination of the aquifer; they universally valued the perception and reality of a clean environment because the Laramie area is their home. Residents of the county who actively participated in CAP Network were concerned about the aquifer being used as a political sword against private property rights and as a means to get rid of rural subdivisions. They were also concerned about hazardous material transport on the nearby interstate highway.

The university was clearly in the middle of the conflict from many different perspectives. University faculty and staff reside in both the urban and rural areas. All rely on the Casper Aquifer for their drinking water supplies. The university also operates and maintains a golf course located near the western boundary of the aquifer protection area, and new residential development associated with expansion of the golf course was proposed for areas overlying the aquifer. After it became clear that the proposed expansion was controversial, the university reverted to the status quo. But the dueling experts problem that emerged concerned the university for the obvious reason that the academies are based on rational discourse about science.

What are your desired outcomes?

On the basis of early and late efforts at aquifer protection, the city remained concerned about septic tanks and fertilizer use in the rural subdivisions overlying the Casper Aquifer. The concerns were not limited to just the nitrates derived from the septic tanks and lawn fertilizer, but also old wells serving as conduits for contaminants to migrate from the surface to the groundwater.

The rural residents associated with CAP Network acknowledged that nitrates are in the Casper Aquifer. But the question first and foremost on their minds was whether the dissolved concentrations of nitrates constituted an actionable problem for the rural or urban residents. Rural residents with CAP Network supported the original aquifer protection program, but they were concerned about the lack of reasons for revising the original Casper Aquifer Protection Area delineation or for the management programs that were developed by the volunteers that were approved by the city and county in 2002. Their rationale was simple: the city requested review of the original aquifer protection program by the WDEQ in 2006. The city's actions became suspect when a request for proposal for an update of aquifer protection program was solicited before receiving WDEQ comment. The rural residents desired an honest dialogue about the geology of the Casper Aquifer, since many of the residents living in the subdivision overlying the aquifer are professional geologists and engineers who work for the many local consulting firms or at the University of Wyoming.

The county advisory board volunteer wanted to see a sewer line installed in the developed areas overlying the Casper Aquifer. Likewise, comprehensive monitoring of the aquifer was needed throughout the aquifer area, as well as for the many water-bearing or subaquifers composing the Casper Aquifer. The volunteer acknowledged the uncertainty associated with well construction and contaminant transport, but desired monitoring of well construction designed to assess the subaquifers. Regular water level and water quality data collection were a must. A centralized database was needed for all well logs, septic systems, well servicing records and septic tank servicing records. It was important to confirm or deny anecdotal reports of questionable septic system construction.

A faculty member at the university echoed the desire to see a monitoring well program installed to better locate 'the problem' nitrate contamination areas. The faculty member also considered the monitoring program along the interstate corridor as a step in the right direction for assessing the chemical integrity of the Casper Aquifer and as an early warning system in case of a spill along the interstate.

The city resident who actively participated in Citizens for Clean Water desired to see meaningful limitations on additional inputs to the aquifer, including, but not limited to, septic tanks, herbicides, pesticides, hydrocarbons associated with housing developments and any waste disposal that could potentially add contaminants to the aquifer. The resident acknowledged that multiple sources could be responsible for the observed nitrate contamination in the rural subdivisions, but that this still represented contaminant loading to the aquifer. The resident also mentioned that the covenants associated with some of the older rural subdivisions indicated a need to hook up to sewers when they became available. In the interim, it was important to slowly upgrade the older existing septic tanks to newer technology. Finally, there was the acknowledgement that installation of sewer lines to rural areas also opened up additional areas within the Casper Aquifer to new residential development. It was important to gauge new development against the preservation of viewsheds, open spaces and planned growth and with fact-based decision making and appropriate regulation in order to preserve community goals. There was a perceived need to stop paralysis by analysis and apply common sense to the issues.

The county commissioner indicated a strong need to protect the aquifer while at the same time having the scientific information to help make decisions on whether the observed concentrations of nitrates were a function of septic tanks or the surface application of fertilizer, and whether the existing well completions were permitting contaminants to make it to the aquifer. The commissioner indicated that the scientific information was needed to stop the 'cherry picking' of data. If nitrates were naturally occurring in the aquifer, that was as important to know as whether wells were serving as conduits. A sewer system had been planned for extension to one of the older rural subdivisions, but the construction of the wastewater treatment plant for the city consumed the funds to act on those plans. There was a real need to get rid of the hostility between the city and county residents.

What is the history of collaboration and the potential for future collaboration?

The county commissioner held office in the 1990s, when the county sponsored the EAC's collaborative efforts and used volunteers to complete the Casper Aquifer Protection Area delineation and the associated supporting sections of contaminant inventories and management plans. The commissioner considered it possible to restart these efforts and also thought the collaborative learning process needed to be revisited every two years. Conversely, the city council member did not express interest in sponsoring additional collaborative learning activities.

The city resident remained willing to consider joint fact-finding efforts and collaboration, but was skeptical, as the situation has made city and county residents increasingly polarized. The resident attempted to sponsor more collaborative learning, since this was a missing component for the aquifer protection process.

The rural residents expressed no interest in joint fact finding, since there is a general lack of trust that emerged after they provided access to wells for sampling, only to have the data 'cherry picked' and used against the rural subdivisions. The city's selection of an out-of-state consultant to update the aquifer protection area was a destructive event. The residents were cautiously optimistic that collaborative learning could be implemented. However, they would support a facilitated dialogue – but not like the science panel held in 2010 because they felt that a public hearing is not the proper forum for the debate on science-based issues associated with the Casper Aquifer.

The joint county advisory board volunteer for environmental advising and planning considered additional meetings on the topic of aquifer protection as 'simply kicking the can down the road'. If collaborative processes were reinitiated, then it would be important to have well-defined goals and schedules. The collaborative efforts would be useful for redefining the western boundary of the aquifer protection area using all of the available well logs. But the volunteer also expressed uncertainty as to whether it would work, as it would more than likely depend on who the participants were in the process. The university faculty member expressed a desire for more joint fact finding and collaborative approaches, but also noted that the atmosphere for collaboration was poisoned and had reached the point of no return in about 2007.

What is the role of the university?

The rural residents indicated that a hydrologist like the university faculty member who once lived in the area no longer existed as a campus resource. They were concerned that university faculty associated with the environmental science program had an agenda, based on the faculty member's participation in past petitions regarding the proposed golf course, the media and communications with the city.

The joint county advisory board volunteer thought that the university could play multiple roles, such as (1) assisting in redefining the western boundary of the aquifer protection area and (2) providing assistance from the law school with mediation and facilitation of meetings. As far as the volunteer could recollect, the university has not been involved since 2010.

The university's past role on the aquifer protection program was educational and included experiential learning for geography students, who used global positioning systems (GPS) and Geographic Information Systems (GIS) to map potential contaminant sources for the original CAPP. The university also once served as a moderator for past public meetings on aquifer protection. However, discussions with a prominent faculty member indicated the university no longer wanted a role in the aquifer protection program because they did not want the publicity. For example, some faculty and staff have used university resources for releasing public position statements, only to have to recall their statements or have them quashed and redefined as personal, not official, opinions.

The city citizen indicated that the university did not have a past role and was disappointed the university was not more involved. The citizen expressed the view that the university does have value in presenting facts, interpreting existing data and playing more of a technical role. The university could also serve in the role of group facilitation.

The county commissioner opined that the university always has a role in the Casper Aquifer protection and referenced the past role of a retired geology professor, who not only lived in the rural subdivision on the Casper Aquifer, but also kept well logs and water levels while monitoring wells located on his property.

What has been the role of the press?

The press is often thought of as an unbiased source of information and as a portal for the exchange of ideas and comments derived from the community through letters to the editor, as well as comments on articles. But the press can also fuel conflict through biased reporting, both perceived and real, of interviews and through editorial review and content.

The rural residents considered the local newspaper to be biased for the city. They reported it was not unusual for their letters to the editor that were inconsistent with city stance regarding aquifer protection to remain unpublished. Press releases about other activities on aquifer protection unrelated to the city's activities would also often go unpublished. Regional news agencies would request interviews with county residents, but the final article would only provide one side of the story. For example, rural residents recalled being interviewed by a reporter on the Casper Aquifer situation, only to later learn that their interviews were not part of the article by Hesse (2012). Likewise, in a letter to the county, one rural resident wrote, 'Frankly it was difficult to determine if the reporter was mongering fear, impending doom for our Casper aquifer or simply looking for someone to support some plan to outlaw septic systems east of Laramie' (Richter 2008).

The county advisory board volunteer indicated that the media reports what they hear. Articles are sometimes dominated by the same individuals at public meetings, but this may be an anomaly. The volunteer considered that the letters to the editor, online comments and anonymous comments contributed to some of the hostility and misinformation about aquifer protection. The university faculty member considered the media to be biased toward the city, and that the bias was because the paper had a more direct line to the city council. The faculty member thought that the media kept things 'stirred up as it sold newspapers'.

One city resident opined that the media was not necessarily all to blame, and that the purchasers of media had a responsibility to urge the media to give an objective airing of public policy issues. In a comment posted on an article in the local newspaper, a different city resident urged the media to provide more objective reporting on local aquifer protection (Baumann, 2013):

> I was at the meeting, along with approximately four other citizens who were in attendance. Apparently the reporter and I were at two entirely different meetings. Because the city learned that there is no risk to the city wells and current trends demonstrate that it would take over a Century to see even slight increases (well below half of maximum contaminant levels), if any increase at all, in city supply wells they have decided that only monitoring is necessary to insure that the report is correct in its findings going forward. The well samples the reporter discusses were those collected years ago by a city staff person who left Laramie years ago and are old news. The methods she used for collection led to errant results and the analysis have been questioned; the couple of wells with increased nitrates were retested by the home owners, and found normal nitrate levels or improperly sealed wells that have been corrected. This article appears to be directed at creating fear and to cause false alarm.

The county commissioner considered the press instrumental in educating the public about aquifer protection. Although there were good, honest reporters, part of the problem was that some people played to the media during public meetings.

Applying the Hydro-Trifecta Framework

Recall that the Hydro-Trifecta Framework is built around the foundation of a compass. The compass case houses skills that serve as the foundation for the sighting mechanism for the modalities of negotiations, which point to the graduated circle of transdisciplinarity in the search for 'new' science and collective action.

Scale targeted skills

As can be seen in the stakeholder assessment summary listed in Table 5.2, the conflict over the Casper Aquifer Protection Area spans many scales. Interpersonal conflicts are common among the scientists and engineers associated with the city and county, as well as those working with the NGOs. Interagency conflicts exist between the city council and county commission and their respective volunteer subcommittees. The intersectoral conflict extends beyond the boundaries of Albany County to the entire state of Wyoming, as different state agencies and legislative bodies are directly and indirectly involved with reviewing and approving the aquifer protection area delineation, providing technical and educational resources and funding resources and institutional resources. The conflict briefly touched upon the interstate scale when an out-of-state university was consulted on one of the proposed desired outcomes regarding a land swap.

Competency targeted skills

Recalling that individuals require a mix of competencies that vary for the same occupation, it is reasonable to assume that current competencies can be enhanced in the spirit of lifelong learning. The competency targeted skills include (1) knowledge and cognitive competence; (2) functional competence; (3) personal or behavioral competence; (4) values and ethical competence; (5) communication competency; (6) cultural competence skills, including understanding the more nuanced definition the urban-rural divide; and (7) online competency including social media, social networking and video.

On the basis of the existing information, it is clear that the affected parties have a high level of knowledge, functional and values competencies. Listening and communication competencies are lifelong learning skills that regularly require updating and refining. Cultural competency could be improved in this situation, as it is clear the urban-rural divide is keeping the parties from better communication. Online competency could enhance cultural competency skills through the use of a jointly developed documentary video.

Program targeted skills

Systems thinking and collaborative learning permit one to see patterns of interdependency. Figure 5.4 is a situation map of the Casper Aquifer protection program. The political landscape for the Casper Aquifer is complex. According to hydropolitics expert 'Tony' Allan, the importance of 'mapping' the terrain that a water expert anticipates navigating cannot be overestimated, since 'Understanding the political landscape into which one wants to introduce new concepts and approaches is more important than being expert in those concepts' (see Blenckner, 2008, p8).

Careful examination of the situation map depicted on Figure 5.4 reveals five overlapping audiences involved in the situation: elected representatives,

Table 5.2 Summary of aquifer protection stakeholder assessment

Group	Connection to aquifer protection	Ideological considerations	Desired outcome	Political considerations
City Council	City drinking water.	Protect drinking water for the greater good.	• Reduce threats to aquifer by old wells, old septic systems, fertilizer use. • Balance business development with hazard reduction. • Sewer line to rural areas.	• Reelection. • Lobby legislature and University of Wyoming.
County Commission	Well water supply for rural residents.	Protect drinking water for rural residents and preserve property rights.	• Protect aquifer. • Stop selective use of data. • Need scientific data that former UW professor collected. • Reduce hostility. • Revisit every two years.	• Reelection. • Lobby legislature and University of Wyoming.
Environmental Advisory Committee	Jointly appointed by city council and county commission.	Advise city council and county commission on issues regarding the environment and planning for the community at large.	• Delineation of protection areas; management plans. • Boundaries for management.	• Volunteers selected from citizens located in urban and rural areas. • Subject to lobbying from many factions.
City Planning Commission/ County Planning and Zoning Commission	Separate entities. Appointed by city council. Appointed by county commission.	Advise respective elected entities (city council or county commission) on issues regarding the environment and planning for the community at large.	• Boundaries for management. • Stop wasting taxpayer dollars on deep wells vs. protecting groundwater.	• Volunteers selected from citizens located in urban and rural areas. • Subject to lobbying from many factions.

Consultants	Geologists and hydrologists delineate protection areas, assess risk, design mitigation.	Professional service, reputation, profit.	Technically and legally defensible boundaries and water quality sampling data.	• Professional licenses. • Contracts with city, county and state.
Citizens for Open Space and Outdoor Recreation (NGO)	• Drinking water. • Some members are UW faculty and staff. • Petition city council against golf course development on aquifer. Petition city council and county commission on use Special Purpose Tax to purchase land.	Drinking water protection, preservation of rural open space and the enhancement of recreational opportunities.	Access to forest from city through hiking and biking trails.	Form coalitions with other NGOs and members of city council and county Commission.
Citizens for Clean Water (NGO)	• Drinking water. • Petition county commission on aquifer protection boundaries. • Some members are UW faculty and staff.	Aquifer protection boundaries. Aquifer contamination.	• Meaningful limitations on additional input to aquifer. • Fact-based decision making.	• Form coalitions with other NGOs and members of city council and county commission. • Lobby university and legislature for land swap/purchase. • Lobby Wyoming Board of Geologists.
Casper Aquifer Protection Network (NGO)	• Homes located on aquifer and supplied by individual wells. • Some members are UW faculty and staff.	Preservation of clean, safe groundwater, now and for generations to come, while retaining our property rights.	• Honest dialogue about geology in protection area. • Groundwater contamination is local issue and can be solved locally.	• Form coalitions with landowners and members of city council and county commission. • Lobby legislature for limits on extra-territorial jurisdiction authority. • Lobby Wyoming Board of Geologists.

(Continued)

Table 5.2 (Continued)

Group	Connection to aquifer protection	Ideological considerations	Desired outcome	Political considerations
Legislature	• Some legislators live in city and county. • Proposed state purchase of 11,000 acres. • Pass HB85 limiting extra-territorial jurisdiction authority	What happens in one region may expand to other regions of the state.	Represent citizens of state.	Reelection, funding limitations, bipartisanship.
University of Wyoming	• Expansion of golf course on aquifer. • Comanage nearby large ranch with other university suggested for land swap. • Mediate science panel. • Assist junior high student with science project.	• Science, engineering and social science expertise. • Laboratories used for water quality analyses.	• Monitoring well program for aquifer and I-80. • Education and outreach.	• Funded by legislature. • Faculty and staff live in both urban and rural settings.
Department of Environmental Quality	• Statewide Wellhead and Aquifer Protection Program. • Review submitted documents.	Oversees the management of Wyoming's natural environment.	• Groundwater protection statewide.	• Funded by legislature.
Wyoming Water Development Commission	Advise governor on proposed water development projects.	Promote the optimal development of the state's human, industrial, mineral, agricultural, water and recreational resources.	• Limit industrial siting. • Rural development not a problem on aquifer.	• Funded by legislature. • Advise water committees and governor.
Wyoming Board of Professional Geologists	Geologic work associated with delineation of protection area.	Protect the health, safety and welfare of the public by promoting the practice of geology by professional geologists.	Geologic work for cities and counties completed by licensed geologists.	Funded by registered geologists in Wyoming.

Junior High Student	Science project on background concentrations of nitrate in aquifer and natural sources of nitrate from plants.	Education.	Science fair award.	Parents reside in rural part of county within aquifer protection area.
Science Panel	Debate the history, science and geography of wellhead and aquifer protection program.	• Multiple working hypotheses. • Junk science unacceptable. • Normative science unacceptable.	Limit scientific debate in the editorial pages and council/commission meetings.	Panelists and mediator derived from consultants, NGOs and academics.
Large Acreage Landowner	• Lands located within designated protection areas. • Provide access to junior high student for science projects.	Returns on financial investments.	Acquire 50,000-acre Y Cross Ranch, co-owned by the University of Wyoming and Colorado State University.	Legal counsel in legislature.
Press	• Articles and radio reports on aquifer protection. • Citizen letters to the editor.	Unbiased reporting.	Sell newspapers and online subscriptions. Increase listenership.	• Editorial board. • Public radio located at University of Wyoming.

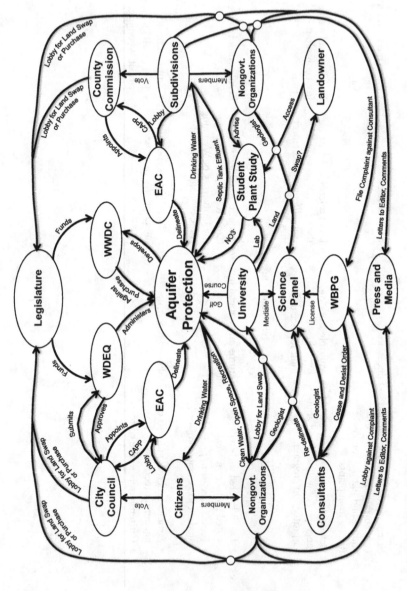

Figure 5.4 Aquifer protection situation map

technical/scientific 'water experts', regulatory agencies, the academies and the public at large. While science remains at the core of aquifer protection issues, the 'types' of science that are involved in this situation are manifold: regulatory or policy science, where research questions are framed by legislators and regulators (WDEQ, WBPG, EAC, county, city); anecdotal, local or 'traditional knowledge' science defined by Ozawa (2005) as information gained by resource users through experience with that resource over time (students, urban residents, rural residents, NGOs, geologists, well drillers, septic tank installers, etc.); advocacy science typical of courtroom situations (landowners, consultants) and academic science typically undertaken by scientists working at government agencies and universities, where the research questions are framed by scientists and driven by rational analysis and expert judgment (Wyoming Water Development Commission [WWDC] and the universities).

The common thread to the different types of sciences is that 'water experts' come in many forms, ranging from scientists and engineers with training or experience in groundwater resources to the middle school student – each expert serving as a catalyst in the conflict, and each seasoned a little differently by their relationship with each other. The map is useful for identifying entry points for moving forward on the path of aquifer protection. It also points out that future collaborative learning on the aquifer protection program could benefit from the use of online video to share messages between individuals who could not attend meetings and consultations either due to schedule or travel, or due to fear of intimidation.

Which siting mechanism should be used?

Water security?

This siting mechanism primarily addresses risk and utilizes a web of climate, energy, food, water and community to define what might be tolerable risk for water use and reuse without getting into 'trouble'. While the value of 'clean' groundwater is universally acknowledged between the city and county residents, the city's consultant determined that the risk of nitrate contamination to the city's groundwater sources is low at this time. The nitrate concentration in the rural residential wells is variable, but there is no universal agreement on what constitutes a 'problem' in rural subdivision, since there are no drinking water standards for private wells.

Water diplomacy?

This siting mechanism acknowledges that water crosses multiple domains, multiparty negotiation and coalitional behavior. It primarily addresses interests and sets its sights on the flexible uses of water and joint fact finding to create value, rather than 'one party wins, one party loses' – also known as the mutual-gains approach to value creation. One of the challenges associated with the Casper

Aquifer protection situation is defining value. Does value include just 'clean' water, open spaces and hiking trails and ridgelines and viewsheds, or does value recognize private property rights?

Water conflict transformation?

Recall that the water conflict transformation 'needle' is used to point disputants between the issues of rights, needs, benefits and equity, while at the same time moving beyond institutions, such as the city and county, toward securing information and creating incentives – all in the quest to create a new superordinate identity. This siting mechanism primarily addresses identity-based conflicts. Identity-based conflicts are relatively intangible and are rooted in the dynamics of history, religion, culture and values. These conflicts continue for a long time. Goals tend to extend beyond advancing concrete interests to include upholding dignity and reputation. The basis of the water conflict transformation process is to address underlying value differences, identity issues or unacknowledged emotions in a different way than disputes over tangible resources. One goal is to create a superordinate identity in order to get all parties to work together. Clearly, the adversarial relationships between the urban and rural residents, coupled with the nearly 15 years of groundwater and aquifer protection efforts to delineate and manage the Casper Aquifer Protection Area, underscore the depth of the identity-based conflict.

The graduated circle of transdisciplinarity

Recall that transdisciplinary partnerships move beyond existing disciplines to new higher-level synthesis – i.e. to 'invent new science', develop systemic thinking, support citizen or 'traditional' science, interpret social and economic implications of science for political purposes and form alliances with other knowledge cultures (Max-Neef, 2005; Brown, 2008; Patterson et al., 2013).

What exists?

Recall that both city and county have much invested in the logic of the aquifer protection program, including the geology and hydrology in order to learn 'what exists'. Citizen science and local knowledge contribute much to what is already known as well.

What are we capable of doing?

The various versions of CAPP reveal that the city, county and both the urban and rural residents can develop win-win situations regarding aquifer protection. Not all of the resolutions necessarily leave everyone happy, but rather the outcomes are something that all parties can live with. When landowners drill exempt wells, they become their own de facto water supply managers. There must be some recognition that rural residential well owners may lack the hydrologic knowledge and expertise to develop a long-term water supply, but there must also be recognition that some rural residential well owner may have the expertise (Bracken, 2010).

What do we want to do?

The efforts at attempting a land swap or land purchase of the lands in and near the aquifer protection area underscore that there is political will to develop long-term solutions to aquifer protection, but it comes at a price.

What must we do?

Looking forward in terms of values, ethics and philosophy, it is clear there needs to be more acknowledgement of local and traditional knowledge. Local knowledge cannot be ignored. Local knowledge comes in many forms, including homebuilders, well drillers, septic tank installers, citizen science and students working on school projects. Volunteers cannot be ignored.

The politicization of science regarding the Casper Aquifer must be eliminated in order to move forward with protecting the aquifer. All parties desire objective science, but, as Pielke (2007, p62) argues, 'expectations for science to resolve political conflict almost always fall short because science provides an "excess of objectivity" useful in support of a broad range of conflicting subjective positions.' Additionally, 'rather than resolving political debate, science often becomes ammunition in partisan squabbling, mobilized selectively by contending sides to bolster their positions. . . . Because science is highly valued as a source of reliable information, disputants look to science to help legitimate their interests. In such cases, the scientific experts on each side of the controversy effectively cancel each other out, and the more powerful political or economic interests prevail, just as they would have without the science'.

More joint fact finding regarding the Casper Aquifer, like that used by the volunteers who completed the 2002 delineation report, is needed. Multiple working hypotheses have been the mantra of the field of hydrogeology for over 100 years, and there is value in rekindling that spirit and passing on this philosophy. The next best place to start is to support a successor effort – the next generation of scientists and engineers who will lead the path forward on aquifer protection.

For example, historic hypotheses regarding nitrate in the Columbia River near Richland, Washington, focused on nitrate and sulfate leaking from an underground storage tank on the Hanford Nuclear Reservation. However, the Hanford Groundwater Analysis Project determined 'that the nitrate came out of the tank. It appears it didn't. It's outside sources. So that's something new' (Boggs, 2013). The Hanford Groundwater Analysis Project is a collaboration between Columbia Basin College and the Washington State Department of Ecology. The conflict cartography of the Casper Aquifer protection program revealed the entry point for the successor effort when a middle school student undertook a study of the soils and nitrate formation in the recharge area of the Casper Aquifer. A transdisciplinary approach may yield new science regarding the issue of nitrates in groundwater, and it might best be handled by students who will be the next generation of aquifer users.

References

Baumann, P. (2013) 'Study shows elevated nitrate levels in Casper Aquifer', *Laramie Boomerang*, 3 May, www.laramieboomerang.com/articles/2013/05/03/news/doc51834045c5b30197169728.txt

Billings Gazette (2012) 'State water chief pans Laramie aquifer purchase', 7 February, http://billingsgazette.com/news/state-and-regional/wyoming/state-water-chief-pans-laramie-aquifer-purchase/article_0d29f750-b254-55b1-892b-a2758fcc8388.html#ixzz2YsEdfivP

Blenckner, S. (2008) 'Do the right thing a little badly . . . ', *Stockholm Water Front*, July, pp8–9

Boggs, M. (2013) 'Hanford groundwater analysis project finds interesting numbers', *NBC Right Now*, 22 April, www.kndu.com/story/22047892/hanford-groundwater-analysis-project

Bracken, N. (2010) 'Exempt well issues in the West', *Environmental Law*, vol 40, pp141–253

Brown, V. A. (2008) *Leonardo's vision: A guide to collective thinking and action*, Sense Publishers, Rotterdam, the Netherlands

Daniels, S. E. and Walker, G. B. (2012) 'Lessons from the trenches: Twenty years of using systems thinking in natural resource, conflict situations', *Systems Research and Behavioral Science*, vol 29, pp104–115 doi:10.1002/sres.2100

Freeze, R. A. (2000) *The environmental pendulum*, University of California Press, Berkeley

Freeze, R.A. and Cherry, J.A. (1979) *Groundwater*, Prentice-Hall, Inc., Englewood Cliffs, NJ

Gyawali, D. (2013) 'Reflecting on the chasm between water punditry and water politics', *Water Alternatives*, vol 6, no 2, pp177–194

Haderlie, C. (2010) 'What's polluting our water? Groundwater panel discusses contamination of the Casper Aquifer', *The Laramie Boomerang*, 7 August, www.laramieboomerang.com/articles/2010/08/07/news/doc4c5cd97e2d088159219270.txt

Hancock, L. (2013) 'Wyoming lawmakers OK curbs on towns' reach outside their borders', *Casper Star Tribune*, 23 February

Hesse, T. (2012) 'Water below: Growing reliance on groundwater may limit activities on overlying surface', *WyoFile*, 12 June, http://wyofile.com/tom-hesse/water-below-growing-reliance-on-groundwater-may-limit-activities-on-overlying-surface/

Huntoon, P. W. and Lundy, D. A. (1979) 'Fracture controlled ground-water circulation in the vicinity of Laramie, Wyoming', *Groundwater*, vol 17, pp463–469

Innes, J. E. and Booher, D. E. (2010) *Planning with complexity: An introduction to collaborative rationality for public policy*, Routledge, New York, NY

Islam, S. and Susskind, L. E. (2013) *Water diplomacy: A negotiated approach to managing complex water networks*, RFF Press, Routledge, New York, NY

Jarvis, W. T. (2010) 'Water wars, war of the well, and guerilla well-fare', *Groundwater*, vol 48, no 3, pp346–350, doi:10.1111/j.1745–6584.2010.00695.x

Livingstone, S., Franz, T. and Guiger, N. (1996) 'Managing ground-water resources using wellhead protection programs', *Geoscience Canada*, vol 22, pp121–128

Margerum, R. D. (2011) 'Beyond consensus: Improving collaborative planning and management', MIT Press, Cambridge, MA

Max-Neef, M. A. (2005) 'Foundations of transdisciplinarity', *Ecological Economics*, vol 53, pp5–16

Moore, L. (2013) *Common ground on hostile turf: Stories from an environmental mediator*, Island Press, Washington, DC

Newman, E. (2012) 'Measuring the cost: City's aquifer protection law balances business development with hazard reduction', *The Laramie Boomerang*, www.laramieboomerang.com/articles/2012/02/11/news/doc4f3600b9f32df344167473.txt

Ozawa, C. (2005) 'Putting science in its place', in J. T. Scholz and B. Stiftel (eds) *Adaptive Governance and Water Conflict: New Institutions for Collaborative Planning*, Resources for the Future, Washington, DC, pp185–195

Patterson, J. J., Lukasiewicz, A., Wallis, P. J., Rubenstein, N., Coffey, B., Gachenga, E. and Lynch, A.J.J. (2013) 'Tapping fresh currents: Fostering early-career researchers in transdisciplinary water governance research', *Water Alternatives*, vol 6, no 2, pp293–312

Pielke, R. A. (2007) *The honest broker: Making sense of science in policy and politics*, Cambridge University Press, Cambridge, UK

Radden, R. (2002) *University of Arizona – cooperative extension watershed resources newsletter*' January–February, vol 1–2K1, no 1, p2, https://cals.arizona.edu/yavapai/newsletters/water/archive/wrfeb2002.pdf

Richter, H. R. (2008) 'Letter to planning and zoning commission, Aquifer Protection Plan' at Albany County Courthouse, Laramie, WY (personal letter)

Rothman, J. (1997) *Resolving identity-based conflict in nations, organizations and communities*, Jossey-Bass, A Wiley Company, San Francisco, CA

Stauffer, G. (2008) 'Finding community interest', *Laramie Boomerang*, 18 July, www.laramieboomerang.com/articles/2008/07/18/news/doc48816be60e96c169175660.txt

Van Vugt, M. (2009) 'Triumph of the commons: Helping the world to share', *New Scientist*, no 2722, pp40–43

Witten, J., Horsley, S., Jeer, S. and Flanagan, E. K. (1995) 'A guide to wellhead protection', American Planning Association Planning Advisory Service Report No. 457–458, American Planning Association, Chicago, IL

Wittman Hydro Planning Associates, Inc. (2008) 'Estimation of Casper Aquifer recharge using the soil-water balance model', Consultant's report prepared for the city of Laramie

WWC Engineering (2013) 'East Laramie wastewater feasibility study', Consultant's report prepared for the city of Laramie

6 A star is born
Documentary film as a tool in conflict resolution

George Bernard Shaw, playwright and 1925 Nobel Laureate in Literature, once said, 'The problem with communication . . . is the illusion that it has been accomplished'. If one wonders why no one is listening to groundwater scientists and engineers, we have ourselves to blame. Communication in the earth sciences has to extend beyond just publishing papers in technical journals. Sophocleous (1997) elucidates that professional communication also goes beyond delivering a journal publication to a manager's desk. It also has to extend beyond relying on PowerPoint presentations to disseminate ideas at technical conferences, where we are doing nothing more than 'preaching to the choir'. Gyawali (2013) avers that if academics in hydrology desire politicians in the villages and districts of the Global South to listen to them, they had better learn to translate their concerns into a language that is understood in local political terms, something they have mostly failed to do. So the purpose of this chapter is to promote the increased use of video to accomplish better communication between individuals negotiating for the myriad uses of water. A few examples of the use of video by students are provided to show the ease of creating videos using the rapidly changing technology that is as easy as using a cellphone.

Ragone (2002, p457) tacitly implies that communication is a key competency to be acquired by indicating that hydrologists will profit by giving 'us a chance to explain to a broader audience why groundwater is central to the well-being of world populations'. Sophocleous (1997, p561) underscores the importance of communication in transferring 'research findings to the field' and conveying 'water-users' needs to the researchers' and suggests that the 'breakdown in communication accounts for the persistence of such misguided concepts' as safe yield. Kendy (2003, p3) suggests that 'successful collaboration requires that we communicate effectively both within and across disciplines' in order to contribute in a meaningful way to groundwater sustainability. And in their seminal work on the 'silent revolution' of intensive groundwater use, Ramon Llamas and Pedro Martínez-Santos (2005a, 2005b, 2005c) echo the importance of communication through educating stakeholders on the fundamental principles of groundwater resources. Yet, Dickerson and Callahan (2006) underscore the challenge associated with communicating about groundwater: because education about groundwater is not a priority in most cultures and countries, even with some direct instruction, much of the common understanding about groundwater involves 'inappropriate mental models'.

Movies, television, billboards, magazines and the Internet flood mass audiences with images that shape our mental models on a broad spectrum of topics and at scales from the individual cell-phone level to the many options for social networking and social media that reach billions. Water and the conflicts associated with its use are popular topics in film. Gleick (2012, 2013) provides an inventory of popular movies and films with a water connection dating back to 1921, where *Three Word Brand* portrayed a skirmish over water rights in Utah. This topic is still in play today between the states of Utah and Nevada over plans to build a pipeline to pump groundwater from a shared aquifer located in northeastern Nevada to Las Vegas. Perhaps the most famous documentary film on water is *Cadillac Desert*, which in 1997 brought to life Marc Reisner's famous tome on the Colorado River (Reisner, 1993). *Cadillac Desert* spawned a new way to 'talk about water' and the genre of water documentaries has expanded dramatically ever since, as video and digital editing software have become easier to use (see Appendix B).

The digital revolution fostering social media also gave voice to short, water-related videos and animations on YouTube (www.youtube.com) and many other video-sharing websites. In her blog, *Water for the Ages*, my colleague Abby Brown compiled a list of 50 videos and film festivals with water and sanitation themes (Brown, 2013). TheWaterChannel (2013) was launched as an open resource in 2009 under a partnership between private industry and UNESCO Institute for Water Education to 'support education and awareness in water by making video material available that is often scattered and easily lost'. In 2013, TheWaterChannel had over 1,700 videos in over 40 categories; 84 videos can be found under the groundwater category alone (TheWaterChannel, 2013).

But the use of video is not limited to small-scale projects. I participated in the Groundwater Working Group of the Global Environment Facility (GEF) funded project IW:Science, formally designated as 'Enhancing the use of science in International Waters projects to improve project results', that was started in 2009 to (1) identify and document science as used in GEF IW projects and (2) analyze the use of science in the selected set of GEF IW projects against the background of relevant aspects and from different angles of view. To get a feel of the importance of the GEF projects on international waters, their portfolio provides $11.5 billion in grants and leveraged $57 billion in cofinancing for over 3,215 projects in over 165 countries. While the Groundwater Working Group examined many core questions, I was assigned the task of evaluating 'Communication of Science Knowledge in Groundwater'. My review of the GEF investments in groundwater showed that concepts 'outside the current paradigm' must be considered, given the breadth of stakeholders and the ensuing negotiations over newly synthesized information regarding the groundwater resources. What I discovered were good examples of the integration of video in groundwater situations within the Guaraní Aquifer project, where TV microprogrammes were produced in Spanish and short videos, such as *The Great Guarani Aquifer* (Gefiwlearn, 2004) and *Formation of the Guarani Aquifer* (Babybell5000, 2004) were professionally produced on the political aspects of the Guarani Aquifer system and the technical aspects of the groundwater system, respectively. Likewise, for the GEF-sponsored Southern Africa Development Community (SADC) project, my summary singled out television, radio, brochures, posters and newspapers as the best media for disseminating information to policy makers, and community meetings as the best means of disseminating to rural communities.

But the larger question is can just anyone make a video that has impact? Scripps Institution of Oceanography at the University of California–San Diego thinks so, and offers summer courses to introduce young scientists to the world of media, blogging and filmmaking – because scientists have been loath to interact with the media. Science journalist Chris Mooney opined in the *Washington Post* that '[s]cientists need not wait for former vice presidents to make hit movies to teach the public about their fields – they must act themselves' (Mooney, 2010). The importance of effective communication in science focuses on negotiating 'mental models', reducing conflict and moving towards collective thinking and action, as 'there can be no negotiation without communication' (Fisher et al., 1981).

The facilitative filmmaker

The new paradigm of the 'compassionate' water resources professional proffered by Berndtsson et al. (2005) concludes that water experts must establish strong linkages with the socioeconomic and human sciences, including how to approach interest groups and decision makers, meet opposition and negotiate, act as educators and trainers and explain methods and visualize techniques in a 'pedagogical manner' to transfer knowledge. Transferring knowledge requires an appreciation for multisensory teaching. While the data presented in Table 6.1 are regularly debated in the field of professional communication, it is clear that people learn more and better from multiple-sense presentations (Fleming and Mills, 1992; Othmana and Amiruddinb, 2010).

We recall that the modern approach to water conflict resolution is to recruit a neutral third party, preferably one who has dual competencies in substance as well as process, who serves as an impartial party and who encourages dialogue (Scher, 1997). Yet the digital revolution has enabled the postmodern approach for rationale discourse regarding water. Collaborative modeling, decision support systems (DSS)

Table 6.1 Presentation styles, senses and percentage of material retained

Presentations and senses	Percent retained
Reading	5–10
Reading plus hearing	10–20
Reading, hearing, plus visual	30
Reading, hearing, visual, plus speaking	70
Reading, hearing, visual, speaking, plus doing	>90

and other variants of 'mediated modeling' are increasingly sophisticated tools for integrating the various aspects of the water sciences and engineering with conflict resolution and negotiation (Chen et al., 2004; van den Belt, 2004). Online dispute resolution is commonly used for e-commerce and is emerging in the field of water conflict resolution, wherein electronic communication serves as the 'fourth party' in the mediation process (Katsh and Rifkin, 2001, p93).

Less well known is the use of documentary film as one means of facilitating dialogue and mutual understanding among stakeholders in a water situation. Can documentary films facilitate cooperative negotiation toward more resilient management of water, especially conflicts over groundwater and aquifers? In trying to describe how online dispute resolution software can act as a neutral third party, Katsh and Rifkin (2001, p91) quote Jim Melamed of www.mediate.com: 'the mediator is someone who is an educator, communicator, facilitator, convener, translator, questioner and clarifier, process advisor, devil's advocate, catalyst, and detail person'. Watson (2012, p9) suggests that a strategically planned, filmed and edited documentary film exploring stakeholder interests and values can satisfy all of those descriptions and can theoretically facilitate mutual understanding and conciliation between stakeholders. Further, she offers that 'since viewers are active and evaluative consumers of media, a film has the potential to assist stakeholders as they appraise and gain deeper understanding of both their own and others' values. For these reasons, documentary films developed by the "facilitative filmmaker" have seemingly great potential as a facilitative tool in multi-stakeholder water-resource cooperative negotiation'.

Let's explore three case studies where 'facilitation through film' was conducted. The first is a groundwater situation in the Pacific Northwest, the second focuses on the Columbia River Treaty between the United States and Canada and the third focuses on the Good Water Neighbors project between Israel, Jordan and Palestine.

Water Before Anything: *Conflict and transformation in Umatilla County*

Groundwater declines approaching 122 m to 152 m in the deep basalt aquifers underlying the Umatilla Basin (Oregon) in the Northwestern United States have occurred over the past 50 years due to intensive exploitation for public drinking water supplies and agricultural irrigation. The deep basalt aquifer is 'shared' by the states of Washington and Oregon, which includes lands ceded by and reserved for the Confederated Tribes of the Umatilla Indian Reservation (CTUIR).

The state of Oregon Water Resources Commission designated Critical Groundwater Areas (CGAs) to conserve the quantity and quality of groundwater for future generations of farmers, starting in 1976 and ending in 1991, thus precluding additional wells for public drinking water and irrigation supplies. Limited federal, state and tribal funding for water resources planning prompted a community-based approach to groundwater management and the exploration of new approaches to manage the sustainability of groundwater and surface water resources. The specific issues that needed to be resolved included (1) how to reverse or recover from the drawdown of groundwater within the state-designated CGAs, (2) how to manage land use outside of urban growth boundaries that may rely exclusively on individual wells tapping the groundwater stored in basalt aquifers and (3) how to maintain flows in rivers and streams hydraulically connected to groundwater stored in basalt aquifers for salmon fisheries important to tribal culture.

Limited federal, state and local funding required experimentation among different resources to synthesize existing socioeconomic and hydrologic information gathered over the course of nearly 20 years for the 20 member Umatilla County Critical Groundwater Task Force (Task Force). The Task Force was appointed by the Planning Commission and the Umatilla County Board of Commissioners to develop and recommend solutions to short- and long-term water quantity issues in Umatilla County, especially within the CGAs. In 2006, graduate students from Oregon State University synthesized data into a 50-page document to serve as a single-text negotiating document that focused on (1) what was known with certainty and would likely not be contested and (2) what was unknown and likely contestable. A place-based, three-dimensional conceptual model of the hydrogeology and groundwater declines was portrayed on brochures, posters and webpages. These documents served as the 'single text' during the early negotiation phases consisting of preparing the future planning documents to maintain trust and expedite review (Jarvis, 2010a).

A documentary film was requested by the Task Force's Outreach Subcommittee and funded by a grant from the Institute for Water and Watersheds through a grant from the U.S. Geological Survey Water Resources Research Act 104(b) program, and supplemented by a grant from the Wild Horse Foundation, the philanthropic arm of the Confederated Tribes of the Umatilla Indian Reservation. The film documents the different perceptions of groundwater issues in the river basin. The film has interviews that chronicle the concerns, hopes and needs of irrigators in the lower basin and of municipal connections and the Confederated Tribes in the upper basin. Water issues are not unique to the Umatilla Basin and it is important to conduct a close examination of communication process and perspectives.

The documentary movie, *Water Before Anything* (see Appendix B), was made by Sarah Lind-Sheldrick, a graduate student pursuing a Master of Arts in English, Rhetoric and Interpersonal Communication. Lind-Sheldrick was the camera operator, the interviewer, the editor and the sound designer on the film project. She had no training or experience in irrigation, and was not a hydrologist or geologist. As filmmaker to the project, she was concerned with the ethical consideration of the relationship between the filmmaker and the subject. It is the intention of the film to provide the community with documentation of the collaborative learning process that led to the successful recommendation and adoption of a plan for ensuring the sustainability of groundwater.

At the beginning of this project Lind-Sheldrick was aware that she faced challenges in being an outsider to the community with a camera. She framed the community interviews for inconsistency rather than consistency, and she purposefully framed subjects off center. She permitted the interviewee to select where the interview would take place. Sometimes she held the camera, and other times she placed the camera on a tripod and did not move it for the entire interview. The inconstancies of image and lighting translate as the visual literacy that makes the documentary appear more 'real', as opposed to feeling overproduced and staged (Lind-Sheldrick, 2009).

Lind-Sheldrick relied on her unemotional involvement in order to present information as it was presented to her during the interviews. She asked everyone the same basic questions: Who are you and what are some of the things that you do in the community? What is the problem? What is the solution? What do you think is the biggest obstacle? What is your hope for the future? She purposely made the questions open in order to give the power to the subject to frame the problem.

The danger in making a documentary of this nature is the expectation of the community to have all voices heard equally. Each represented group measured the documentary by their own perspective on the groundwater issues. The intent of the documentary was to begin the conversation on local groundwater issues. Although nearly 27 hours of footage were captured, the goal was to keep the documentary under 40 minutes so it could be used in a workshop or classroom setting that would allow time and room for discussion. *Water Before Anything* was eventually used to bring awareness of the field to decision makers and was shown before voting took place on the only water-supply investment made in Oregon during 2008 and 2009. Over 400 copies of *Water Before Anything* were distributed as DVDs to addresses across the globe, before the film was posted on TheWater-Channel, where over 10,000 viewings have been reported.

A River Loved: A Film About the Columbia River and the People Invested in Its Future

The Columbia River Basin encompasses 673,138 km^2 across seven U.S. states and the Canadian province of British Columbia. The river historically hosted a robust salmon fishery and its banks provided trading places for Native American tribes, most notably at Celilo Falls, as described in the documentary *Celilo Falls and the Remaking of the Columbia River* (see Appendix B). During the 20th century, the Columbia was transformed by locks and dams created to harness the river for human needs, as shown by documentary filmmaker Chris Swain in *Source to Sea: The Columbia River Swim*.

As summarized by Watson (2012), the International Joint Commission, created by the U.S. and Canadian governments, convened in the 1940s to discuss flood control and hydropower on the Columbia River. Formal negotiations for a Columbia River Treaty began in 1961. The subsequent agreement provided the United States with 60 years of flood control via storage in Canada, as well as controlled flow and hydropower benefits. In return, Canada received three U.S.-financed dams, $65 million for the 60 years of flood control and one half of the increase in U.S. hydropower revenues.

Reexamination of the Columbia River Treaty was undertaken because the purchased flood control outlined in the treaty ends in 2024. Basin stakeholders and legitimate sovereign entities (i.e., Tribes and First Nations) were not included in the original negotiations and they now desire a seat at the table. Likewise, many contemporary values for the river are not reflected in the 1964 treaty, either due to omission or due to shifting values and perspectives over the last several decades (Watson, 2012).

As part of a graduate student project to support the Universities Consortium Symposium on Columbia River Governance and to build upon the work of Lind-Sheldrick (2009), Julie Elkins Watson (2012) proposed an intervention strategy, 'learning through film', based on the idea of collaborative learning, as developed by Daniels and Walker (2001). The end result was Watson's film, *A River Loved: A Film About the Columbia River and the People Invested in Its Future*, a documentary film used as a medium through which parties could use systems thinking to learn about one another's interests and values, as they gain an enhanced understanding of the Columbia River. With approximately 12–14 hours of time invested for the 37-minute video, Watson attempted to answer the following questions: (1) Can documentary films facilitate cooperative negotiation towards more resilient management of social-ecological systems? (2) Can facilitative documentary films promote dialogue? (3) Can documentary films facilitate understanding/empathy amongst parties? And (4) Can documentary films encourage parties to consider new scenarios?

Like Lind-Sheldrick (2009), Watson (2012) employed the facilitative technique of asking directed, open-ended questions. However, unlike Lind-Sheldrick, Watson held the interviews outside, rather than permitting the interviewees to select the location. She asked stakeholders to share personal, historical, cultural and faith-based stories, rather than focusing only on the textbook information about the treaty. She also focused on highlighting each interviewee's unique ethos, which gave the speaker credibility through history, experience or commitment. She also incorporated logos, the facts and logical arguments each speaker used to support her or his point. Watson argues that credibility and logic can build respect, but cultivating empathy requires a third element: pathos. She wanted the audience to connect with each speaker at a deeper level, and to do this, she needed to create an emotional appeal.

Acknowledging the convenience factor of video, both filmmakers recognized that 'if one cannot bring the decision-makers to the field, one must bring the field to the decision-makers' (Watson, 2012, p28). A virtual fieldtrip, documentary video permits the viewer to 'travel' to the various ends of the basin, see and hear the voices of the stakeholders and more importantly, hear their stories, values and interests uninterrupted. Watson (2012, p44) argues that documentary film 'can give voice to historically disadvantaged parties while also allowing the agency heads to show their more human side, thus helping stakeholders to see and understand one another better at the same time that they learn about more complex substantive matter'.

The results of Watson's study provide good data that documentary video enhances a better understanding of interests and values, better dialogue and new scenario development amongst stakeholders in the basin. She reports that 'over 90% of respondents said that they understood others' perspectives better after watching the film, and about the same amount claimed they thought others who watched the film would understand their personal perspective better' (Watson, 2012, p46).

A Regional Love Story – Friends of the Earth, Middle East

It is beyond the scope of this paper to begin to describe the political tensions associated with groundwater development along the banks of the Jordan River, although the interested reader is referred to Zeitoun et al. (2009) for a review of the situation. During September 2012, two important Middle East water conferences were convened to begin the process of rehabilitating the Lower Jordan River. The first Sustainable Water Integrated Management (SWIM) Transboundary NGO Master Planning of the Lower Jordan River Basin was convened during the Jordan River Stockholm International Water Institute (SIWI) Middle East Seminar, and brought together approximately 80 high-level officials from the region and from the international community. Representatives of all key Jordanian, Palestinian and Israeli decision-making bodies and ministries relevant for the rehabilitation of the Lower Jordan River were present. The second conference initiated a celebration of 10 years of the Good Water Neighbors Project sponsored by the Friends of the Earth, Middle East (FoEME); approximately 250 people from across the region and globe attended the two-day conference in Jericho, Palestine. It was reportedly the first large water conference hosted by the Palestinians. It was clear that just about any option for water was being considered, not only to reduce the potential suffering of thirsty growing population, but also to meet the needs for irrigated agriculture, to continue restoring the green corridor and to save the Dead Sea.

From the perspective of a spectator, both conferences were attempts to highlight community-based problem solving on water issues. Collaboration between entities within the region used new tools to enhance the notion of a potential superordinate identity, 'we are all in this together'. The new tools included the use of original music, song and a jointly produced five-minute video titled *A Regional Love Story – Friends of the Earth, Middle East* (see Appendix B). Both meetings highlighted a shared vision on the preservation of the Dead Sea and other important legacy sites. A large reason to promote stakeholder involvement through documentaries consists of the social learning elements. Social learning can be summarized in one phrase as 'learning together to manage together' (WWAP, 2012, p596).

Systems thinking, situation mapping, network analysis and documentaries

Students and practitioners desiring opportunities to practice systems thinking and situation mapping of 'wicked problems' important to building transdisciplinary programmatic and negotiating skills under the many water negotiation frameworks will find the videos listed in Appendix B useful. A cursory review of the films listed in Appendix B reveals that water is part of conflicts and negotiations for problems relating to water rights, mining, land development, energy development and dueling expert situations across the globe. *Wind River* and *Upstream-Downstream* are two examples frequently used in trainings conducted both in the United States and abroad, as they are relatively short (less than 30 minutes in length) and provide an international perspective on water conflicts, and students are able to develop situation maps while the videos are shown. Each student can be encouraged to share his or her situation map with others, either in small-group discussions or in a plenary session.

Dueling documentaries debating water issues

Building upon the notion of documentary film serving as a medium for discourse, or 'the fourth party', a relatively new phenomenon is emerging with documentaries that serve as an example of 'the devil's advocate' in facilitative filmmaking. For example, bottled water is a $100 billion per year industry worldwide. Many documentaries that focus on the story of bottled water examine the industry's impact on the environment in using groundwater as part of the supply chain, the fossil fuels consumed in manufacturing the plastic bottles and the impact of disposing of the bottles. *Tapped* and *FLOW: For the Love of Water* are two such documentaries that portray the bottled water industry as bad for the environment or as part of efforts to privatize water resources. *FLOW* takes a relatively aggressive stance on a bottling plant located in Michigan that is operated by Nestle Waters, North America, proclaiming that wetlands and springs were dried up due to the operations of the plant. However, Nestle counters *FLOW* with a video of their own that rebuts many of the claims made in the documentary.

Likewise, the documentary *Gasland* portrays the global efforts of hydraulic fracturing for natural gas (fracking) in unconventional shale reservoirs as contaminating groundwater and impacting private wells wherever the gas industry implements the technology. The documentary *FrackNation* was made to counter many of the assertions in *Gasland*. The video discourse between the two documentarians continues with the release of *Gasland, Part II* in 2013.

References

Babybell5000 (2004) *Formation of the Guarani Aquifer* [video file], retrieved from www.youtube.com/watch?v=SnJ_WPtfmfo

Berndtsson, R., Falkenmark, M., Lindh, G., Bahri, A. and Jinno, K. (2005) 'Educating the compassionate water engineer – a remedy to avoid future water management failures', *Hydrological Sciences*, vol 50, no 1, pp7–16

Brown, A. (2013) 'H2O film festivals', *Water for the Ages* blog, http://waterfortheages.org/water-films

Chen, C. W., Herr, J. and Weintraub, L. (2004) 'Decision support system for stakeholder involvement', *Journal of Environmental Engineering*, vol 130, no 6, pp714–721

Daniels, S. E. and Walker, G. B. (2001) *Working through environmental conflict: The collaborative learning approach*, Prager, Westport, CT

Dickerson, D. and Callahan, T. (2006) 'Ground water is not an educational priority', *Groundwater*, vol 44, no 3, p323

Fisher, R., Ury, W. and Patton, B. (1981) *Getting to yes*, Houghton Mifflin, New York, NY

Fleming, N. D. and Mills, C. (1992) 'Not just another inventory, rather a catalyst for reflection', *To Improve the Academy*, vol 11, pp137–155

Gefiwlearn (2004) *The Great Guarani Aquifer* [video file], retrieved from http://youtube/3LpOxQg5AcY

Gleick, P. H. (2012) 'Water in the movies', in *The World's Water: The Biennial Report on Freshwater Resources*, vol 7, Island Press, Washington, DC, pp171–174

Gleick, P. (2013) 'Significant figures', *ScienceBlogs*, http://scienceblogs.com/significantfigures/

Gyawali, D. (2013) 'Reflecting on the chasm between water punditry and water politics', *Water Alternatives*, vol 6, no 2, pp177–194

Jarvis, W. T. (2010a) 'Community-based approaches to conflict management: groundwater in the Umatilla Basin' in A. Earle, A. Jägerskog and J. Öjendal (eds) *Transboundary Water Management: Principles and Practice*, Earthscan, London, UK, pp225–227

Jarvis, W. T. (2010b) 'Water wars, war of the well, and guerilla well-fare', *Groundwater*, vol 48, no 3, pp346–350, doi:10.1111/j.1745–6584.2010.00695.x

Katsh, E. and Rifkin, J. (2001) *Online dispute resolution: Resolving conflicts in cyberspace*, Jossey-Bass, A Wiley Company, San Francisco, CA

Kendy, E. (2003) 'The false promise of sustainable pumping rates', *Groundwater*, vol 41, no 1, pp1–4

Lind-Sheldrick, S. A. (2009) 'Water before anything: Annual report for US Geological Survey', Institute for Water and Watersheds, Corvallis, OR

Llamas, M. R. and Martínez-Santos, P. (2005a) 'Ethical issues in relation to intensive groundwater use', in A. Sahuquillo, J. Capilla, L. Martinez-Cortina and X. Sanchez-Vila (eds) *Groundwater Intensive Use*, International Association of Hydrogeologists Selected Papers No. 7, A. A. Balkema Publishers, Leiden, the Netherlands, pp3–22

Llamas, M. R. and Martínez-Santos, P. (2005b) 'Intensive groundwater use: Silent revolution and potential source of social conflicts', *American Society of Civil Engineers Journal of Water Resources Planning and Management*, vol 131, no 4, pp337–341

Llamas, M. R. and Martínez,-Santos, P. (2005c) 'The silent revolution of intensive ground water use: Pros and cons', *Groundwater*, vol 43, no 2, p161

Mooney, C. (2010) 'On issues like global warming and evolution, scientists need to speak up', *The Washington Post*, 3 January, pB02, www.washingtonpost.com/wp-dyn/content/article/2009/12/31/AR2009123101155_pf.html

Othmana, N. and Amiruddinb, M. H. (2010) 'Different perspectives of learning styles from VARK model', *Procedia Social and Behavioral Sciences*, vol 7, pp652–660

Ragone, S. E. (2002) 'Corporate hydrology: Profit by growth or by acquisition?' *Groundwater*, vol 40, no 5, p457

Reisner, M. (1993) *Cadillac desert*, Penguin Books, New York, NY

Scher, E. (1997) 'Using technical experts in complex environmental disputes', Mediate.com, www.mediate.com/articles/expertsC.cfm

Sophocleous, M. (1997) 'Managing water resources systems: Why "safe yield" is not sustainable', *Groundwater*, vol 35, no 4, p561

TheWaterChannel (2013) 'About us', www.thewaterchannel.tv/en

Van den Belt, M. (2004) *Mediated modeling: A systems dynamics approach to environmental consensus building*, Island Press, Washington, DC

Watson, J. E. (2012) 'A river loved: Facilitating cooperative negotiation of transboundary water resource management in the Columbia River Basin through documentary film', unpublished MS thesis, Oregon State University, Corvallis, OR

WWAP (World Water Assessment Programme) (2012) 'The United Nations World Water Development Report 4: Managing water under uncertainty and risk', UNESCO, Paris

Zeitoun, M., Messerschmid, C. and Attili, S. (2009) 'Asymmetric abstraction and allocation: The Israeli-Palestinian water pumping record' *Groundwater*, vol 47, no 1, pp146–160

7 Peak oil meets peak water
The Disi Aquifer

> Unitization certainly did work in the oil and gas context.
> While it was fought by some, it has proven to be the savior of all.
> – S. E. Clyde (2011)

On July 13, 2013, the Kingdom of Jordan commenced test pumping of 55 wells installed to extract groundwater stored in the Disi Aquifer for the 325-km-long journey in the Disi Water Conveyance Project to Amman, Jordan (Figure 7.1). The project is ambitious; with an anticipated planned extraction of 100 million m^3 per year, it closely rivals the 140 million m^3 per year extraction projected for the Southern Nevada Water Authority groundwater development scheme, with its 423-km pipeline in the western United States. Like the Paleozoic carbonate aquifers shared by the states of Utah and Nevada, the Disi Aquifer is also a Paleozoic carbonate aquifer; it comprises part of the Ram Group shared with Saudi Arabia (Kessler et al., 2000). The Disi Aquifer is a nonrenewable or fossil aquifer with groundwater age dated at 30,000 years (Puri and Aureli, 2009). And like the Utah and Nevada shared aquifer system, the Disi Aquifer does not have a bilateral agreement between Jordan and Saudi Arabia governing groundwater withdrawals. The project was important to Jordan, and it was unencumbered by the veto of neighboring countries (Ferragina and Greco, 2008).

The Disi Aquifer was discovered in 1969. Saudi Arabia has pumped groundwater from the Disi Aquifer since the 1970s for wheat production. Jordan used the pumped groundwater to supply water to the city of Aqaba and the surrounding region, but in the 1980s awarded concessions to large agro-businesses, who pumped without limits. Neither country reports the quantity of extracted groundwater to the other (Ferragina and Greco, 2008).

Contested expertise, dueling experts and multiple working hypotheses

As with most large groundwater withdrawal projects, developing the Disi Aquifer has seen its share of contested expertise. Conflict emerged regarding the wellfield location. The current locations of the 55 wells are not in the same area recommended on the basis of well yield and water quality when the feasibility study

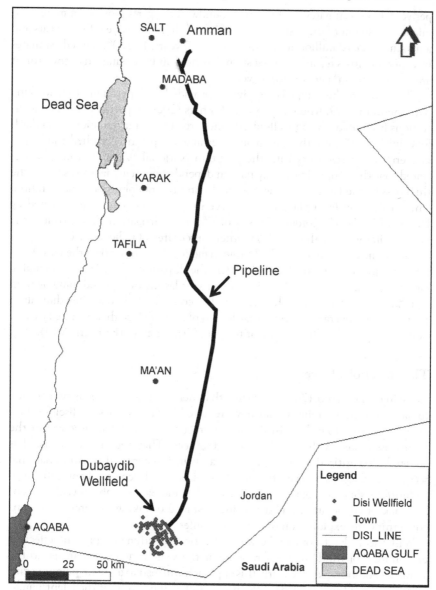

Figure 7.1 Map of Disi wellfield and pipeline

was completed in 1994. The revised well sites were not based on geology and hydrology, but rather on economics.

Ferragina and Greco (2008) report that the sustainable yield of the Disi Aquifer has also served as a source of dueling experts. Early studies on the renewability of the Disi Aquifer were reported at rates of exploitation of 80 to 90 million m³ per year. Later estimates by the same groundwater experts provided a more optimistic assessment of the sustainability of 100 million m^3 per year for 40 years and a maximum of 70 million m^3 per year over 100 years. In 2008, revised estimates by other scientists regarding the sustainable yield and the duration of the project are 100 million m^3 per year for 50 years.

The quality of the pumped groundwater provided the basis for multiple working hypotheses in 2009, when Vengosh and others (2009) reported radium concentrations from 36 of the 55 wells at 20 times greater than international standards for drinking water. But the Jordanian government apparently brushed aside those concerns when scientists with the Egyptian National Water Research Center stated that the contamination by radium depended on many factors, such as the thickness of the layer of sedimentary rocks and the radioisotope concentration, which can differ from place to place, even in the same aquifer, as reported by Lasheen (2013). The Jordan Ministry of Water and Irrigation plans to dilute the Disi Aquifer water with water from other sources free of radioactivity.

Clearly, media coverage has had a role in fueling the conflict over the Disi Aquifer. Ferragina and Greco (2008) chronicle the disparities in anticipated pumping rates ranging from 50 to 200 million m^3 per year lasting over periods ranging from 50 to 200 years. And now with water with low concentrations of radium difficult to secure, media coverage indicates that Jordan plans to reduce their planned extraction rate from 100 million m^3 a year to about 70 million m^3 (Lasheen, 2013).

The power of silence

Ferragina and Greco (2008) indicate that negotiating an agreement between Jordan and Saudi Arabia would have required both countries to disclose their extraction rates from the Disi Aquifer, a practice that is apparently against the norms associated with use of natural resources. They report that Jordan has adopted a 'securitization' strategy, where an issue is securitized as part of national security as a means to keep the issue out of political debates. They call this a 'voluntarily silencing' of the issue. Both Jordan and Saudi Arabia revealed a general lack of cooperation and a continued stance of taking unilateral decisions concerning the exploitation of the Disi Aquifer.

Will the lack of contact between the two governments prevent effective exploitation of the Disi Aquifer, while at the same time cause damage to the storativity of the aquifer? Clearly this approach will lead to a 'silent' pumping race (Ferragina and Greco, 2008). Yet, these scholars indicate that the silent pumping race between Jordan and Saudi Arabia is unlikely to lead to an open conflict. Although it is apparent that the history of the Disi project has been marked by a lack of transparency, there are some cracks beginning to form in this foundation. A 'memorandum of understanding' between the two countries was apparently signed in 2007, where both sides agreed on the following:

1   To keep a reserved area between the two neighboring countries.
2   Not to expand any new agricultural activity in the reserved area.

3   The Jordanian parties implement procedures to limit extraction of ground-water from Saq/Disi aquifer used for agricultural activities existing within the reserved area.

4   Necessity to find a mechanism for cooperation and follow-up of water system of Saq/Disi aquifer, in Dubeideb area in Jordan and Tabouk central water project in Saudi Arabia.

Mechanism for cooperation – the Hydro-Trifecta Framework

The Hydro-Trifecta Framework provides a lens for examining the negotiations and renegotiations for the development of shared aquifer systems. We recall that the Hydro-Trifecta Framework is built around the foundation of a compass case of skills. These skills serve as the foundation for the sighting mechanism for the modalities of negotiations pointing to the graduated circle of transdisciplinarity. While development of the Disi Aquifer is very political as described by Ferragina and Greco (2008), in the following case study I will show that the Disi Aquifer situation is an ideal example of both wicked groundwater and aquifer problems that could implement a transdisciplinary approach for future water negotiations.

Scale targeted skills

It is clear that the scale of the Disi Aquifer situation spans nearly all scales of conflict: intrapersonal, interpersonal, intersectoral, interagency, interstate and international. Each scale has a unique embedded logic. And each scale relies on the outcome of actions by the other scales within the conflict. For example, water consumption at the intrapersonal scale depends on the outcome of the dueling expert situations on water quantity and water quality discussed at the interpersonal scale. Irrigation water supplies in the Disi region depend on the needs of municipal water supplies in Amman. International funding for the Disi Water Conveyance project depended, in part, on the silent approval or disapproval of neighboring Saudi Arabia. How a project is framed, however, such as how the water needs in the Disi region are connected to the regional Red Sea–Dead Sea Water Conveyance Study described by Allan, Malkawi and Tsur (2012), dictates how the international conflict might be handled by obvious, and not so obvious, financiers.

Competency targeted skills

Scientists, engineers, economists and legal scholars with high levels of knowledge, functioning and professional competencies have been involved with the Disi Aquifer development and water conveyance project from the discovery of the aquifer, through its use for irrigated agriculture by both Jordan and Saudi Arabia, to the development of the concept to deliver the pumped water to Amman and to dealing with potential issues of radioactive constituents in the water used for human consumption. It is also clear from one of the most innovative animated videos I have seen that regional projects have discovered the importance of online competency. The video is *The Valley of Peace Initiative* (Gertner Architects' Projects, 2009) associated with the Red Sea–Dead Sea Water Conveyance Study.

What is missing is a larger view of the 'problems', that is, to learn from neighbors with competencies in comparable situations. For example, the United Arab Emirates is quickly becoming the 'go to' source for competencies in aquifer storage and recovery (ASR), particularly as it applies to the use of desalinated water as a supply for storage or 'dilution'. Both Egypt and Yemen have experience with adaptive hybrid instruments for sharing oil and gas, as well as with the redetermination process associated with reassessing the storage and recoverability of fluids through unitization (Weaver and Asmus, 2006).

Program targeted skills

The program development defined in the early stages as described by Ferragina and Greco (2008) suggested a classical 'Tech-Reg' approach, where lenders, institutions and agencies invite, inform and then ignore the parties, as described by Daniels and Walker (2001). Although situation mapping of the Disi project may initially be considered a particular worldview for Jordan or Saudi Arabia, it can also become a shared worldview map as well (Daniels and Walker, 2012). Certainly the project is complex, but the challenge with this project is not to dwell on the myriad details that constitute the whole, but rather with the 'big picture'. Collaborative learning, as described by Daniels and Walker (2001), supplemented by a general morphological analysis, may clearly depict various linkages, thus acknowledging the 'silent' voices and benefitting the process of future decision making in this regard. General morphological analysis is the non-quantified problem-structuring method that builds inference models representing the total problem space and potential solution concepts to the given problem, as described by Ritchey (2013). Some of the collaborative learning could also benefit from the use of online video to share messages between entities who cannot attend meetings and consultations due to schedule or travel, fear of intimidation or politics.

Which siting mechanism should be used?

Water security?

Water security utilizes a web of climate, energy, food, water and community to define what might be tolerable risk for water use and reuse without getting into 'trouble'. The Disi Aquifer situation is already a water security situation for Jordan, as it has adopted a 'securitization' strategy, that is, the project falls within the sphere of national security and out of the political debate (Ferragina and Greco, 2008).

Trouble for both Jordan and Saudi Arabia is the depletion of the Disi Aquifer, with concomitant destruction of the aquifer storage. Food production for Saudi Arabia and drinking water supplies for Amman are at risk. Likewise, the financial risk for the project is already quite large, with nearly $1 billion invested in construction costs. It is clear that additional investment will need to be made to accommodate the source of 'dilution' water to reduce the naturally occurring radioactivity in the produced water, as well as for still-to-be-identified potential water-treatment options.

Another problem for the Disi Aquifer project is energy costs, due to the 250 m pumping lift between the wellfield and Amman, Jordan (Ferragina and Greco, 2008). Additional risks include the high percentage of water losses in the Amman distribution network, which are considered large enough to invalidate the benefits from the Disi Aquifer. And there are risks associated with interference from other uses, specifically the farms overlying the aquifer that use groundwater for irrigation. Ferragina and Greco (2008, p456) cite a prominent Jordanian scientist, who stated that with interfering uses, 'We might end up losing everything.'

Long pipelines tied to wellfields are at risk of damage from conflicts. Gleick and Heberger (2012) chronicle damage to conveyance structures resulting from conflicts that date back to the 1907 bombing of the pipeline from the Owens Valley to Los Angeles, as documented in the video *Cadillac Desert* (see Appendix B). At least three acts of terrorism and war have damaged pipelines in the region since 1965. And regional politics such as the Arab Spring provide little confidence for long-term water security.

Water diplomacy?

Again, we recall that the water diplomacy framework sets its sights on the flexible uses of water and joint fact finding to create value, rather than zero-sum thinking, through a loop of societal, political and natural networks. Neither Jordan nor Saudi Arabia is willing to share data with each other, so the value of 'zero' to both countries remains unknown.

Water conflict transformation?

The water conflict transformation 'needle' is used to help disputants become better listeners and communicators. However, the general lack of contact between Jordan and Saudi Arabia leaves little chance for this path-setting measure at the present time. However, the results of the Arab Spring may provide an opportunity for a new superordinate regional identity, perhaps led by the Arab Water Academy (2013), to build upon for more regional cooperation over groundwater resources.

The graduated circle of transdisciplinarity

What exists?

Both countries have much invested in the logic of the project, including extensive drilling and aquifer testing to learn more about the geology, hydrology, ecology and numerical modeling of the Disi Aquifer groundwater system; in other words, to learn 'what exists'. The problems associated with the Disi Aquifer are still ill defined, however, despite years of study by geologists, hydrologists, engineers, political scientists and financiers. The problems with the groundwater development project are ambiguous and are associated with strong moral, political and professional issues. And there is still little consensus about what the problems are, let alone how to deal with them. In short, the wicked problems associated with the development Disi Aquifer 'won't keep still' long enough to resolve any of them (Ritchey, 2013). Likewise, from a financial perspective, what is also known is that the Disi project did not receive a World Bank investment or other international cooperation funding under the World Bank's safeguard policy because of the lack of 'non-objection' by Saudi Arabia (Ferragina and Greco, 2008).

The 'race to the pumps' – resulting in competitive drilling and production with consequent economic waste, physical damage to the pool and reservoir and with each owner drilling more and pumping faster than his or her neighbor – is not a new problem. In fact, the problem was so prevalent during the early days of many oil fields that the 'unitization' concept was designed in order to permit reservoir engineers to plan the operation of an oil 'pool'. The oil pool helped to conserve petroleum resources and slow the process of 'peak oil', whereby intensive exploitation of petroleum 'reservoirs' led to premature depletion and, in some cases, irreversible damage to the storage characteristics of oil and gas reservoirs. Unitization started in the United States; however, it has spread to other parts of the world and is employed in the Middle East region in Egypt and Yemen (Weaver and Asmus, 2006). Unitization of oil fields was developed to protect the 'corresponding rights' or 'sovereignty' of all pore space owners in the unit and to not waste valuable pore space through joint operation. No case studies exist regarding unitization of aquifer systems, but the principles of unitization are being applied in a few settings.

What are we capable of doing?

The technological aspects of the project in terms of engineering clearly answer the question of 'what are we capable of doing'. Despite the reported problems of depleting groundwater stored in aquifers through intensive exploitation, the aquifer storage space once filled with groundwater nevertheless has great value in alleviating water scarcity. The United Arab Emirates (UAE) have invested $500 million to refill depleted and saline aquifers with 26 million m^3 of desalinated water, a quantity of water sufficient to provide a 90-day supply of drinking water (Henzell, 2012). The UAE consider this investment just one means of dealing with their water scarcity situation. Perhaps UAE can share their knowledge and technology with the Jordanians and Saudis as part of an emerging role in groundwater niche diplomacy.

What do we want to do?

The question of 'what do we want to do' may be addressed by developing a confidential shared-water 'unitization' agreement designed to protect the information, yet still be monitored by an expert who is also a neutral third party. According to

Worthington (2011), the cornerstone of the unitization process is the unitization and unit operating agreement. These are relatively straightforward documents and model forms developed by the Association of International Petroleum Negotiators, and they can be purchased and downloaded for a modest cost. As is common practice in the unitization process, dispute prevention is desired more than conflict resolution; disputes regarding the unitization process are resolved by engaging a 'redetermination expert'. Given that many future conflicts in the Middle East and internationally will occur because of limitations in availability and challenges in the shared use of water resources, perhaps the Arab Water Academy would be willing to fill an important role in groundwater niche diplomacy by hosting a regional water mediation center that houses a neutral third-party unitization redetermination expert.

What must we do?

But the question of 'what must we do' in terms of ethics, philosophy and theology remains problematic. In terms of the Disi Aquifer, all aquifers users might consider preserving aquifer storage for later uses after 'peak water' has been achieved.

Gleick and Palaniappan (2010) note that peak nonrenewable water is groundwater stored in aquifers. But concern over peak water is not limited to only mining nonrenewable groundwater. Narasimhan (2009) indicates that the definition includes aquifers where the storage characteristics have been permanently changed due to pumping. This phenomenon is limited to confined aquifers like the Disi Aquifer.

In my work promoting the unitization of aquifers, the core principles, or '4P' framework, behind unitization of transboundary aquifers are as follows:

1 Promote groundwater exploration and development in underutilized areas, for example, in 'megawatersheds' that are being promoted as a new exploration paradigm.
2 Preserve the storativity of aquifers by promoting local control of groundwater development.
3 Promote private and global cooperation investments in ASR and managed recharge (similar to secondary and tertiary recovery operations used in the oil and gas industry).
4 Prevent disputes by 'blurring the boundaries', thus creating a new community of users with a superordinate identity, for example, 'We are all Arabs' who agree on how to 'share' the groundwater and aquifer storage space.

References

Allan, J. A., Malkawi, A. I. H. and Tsur, Y. (2012) 'Preliminary draft report, Red Sea–Dead Sea Water Conveyance Study Program: Study of alternatives', http://siteresources.worldbank.org/INTREDSEADEADSEA/Resources/Study_of_Alternatives_Report_EN.pdf

Arab Water Academy (2013) www.awacademy.ae/

Clyde, S. E. (2011) 'Beneficial use in times of shortage: Respecting historic water rights while encouraging efficient use and conservation', *The Water Report*, vol 83, pp1–13

Daniels, S. E. and Walker, G. B. (2001) *Working through environmental conflict: The collaborative learning approach*, Prager, Westport, CT

Daniels, S. E. and Walker, G. B. (2012) 'Lessons from the trenches: Twenty years of using systems thinking in natural resource, conflict situations', *Systems Research and Behavioral Science*, vol 29, pp104–115 doi:10.1002/sres.2100

Ferragina, E. G. and Greco, F. (2008) 'Disi project: An internal/external analysis', *Water International*, vol 33, no 4, pp451–463

Gertner Architects' Projects (2009) *The valley of peace initiative* [video file], associated with the Red Sea–Dead Sea Water Conveyance Study, www.youtube.com/watch?v=E1H3SE9xUIg (uploaded on 27 October 2009)

Gleick, P. H. and Heberger, M. (2012) 'Water conflict chronology', in *The World's Water: The Biennial Report on Freshwater Resources*, vol 7, Island Press, Washington, DC, pp175–205

Gleick, P. H. and Palaniappan, M. (2010) 'Peak water: Conceptual and practical limits to freshwater withdrawal and use', *Proceedings of the National Academy of Sciences (PNAS)*, vol 107, no 25, pp11155–11162, www.pacinst.org/press_center/press_releases/peak_water_pnas.pdf

Henzell, J. (2012) 'Ensuring the security of water, "a strategic commodity on par with oil"', *The National*, 30 June, www.thenational.ae/news/uae-news/environment/ensuring-the-security-of-water-a-strategic-commodity-on-par-with-oil

Kessler, S., El-Naser, H., Kawash, F. (2000) 'Temporal trends for water-resources data in areas of Israeli, Jordanian, and Palestinian Interest', U.S. Geological Survey, Executive Action Team (EXACT), Middle East Water Data Banks Project, www.exact-me.org/trends/TrendsReport.pdf

Lasheen, N. (2013) 'Jordan readies the taps on controversial water project', *SciDev.Net*, www.scidev.net/global/water/feature/jordan-readies-the-taps-on-controversial-water-project.html

Narasimhan, T.N. (2009) 'Groundwater: from mystery to management', *Environmental Research Letters*, vol 4, no 3, doi:10.1088/1748–9326/4/3/035002

Puri , S. and Aureli, A. (eds) (2009) 'Atlas of transboundary aquifers, global maps, regional cooperation and local inventories', UNESCO-IHP ISARM Programme, International Hydrological Programme Division of Water Sciences, United Nations Educational, Scientific and Cultural Organization, Paris, www.isarm.net/publications/322, last accessed November 2009

Ritchey, T. (2013) 'Wicked problems: Modelling social messes with morphological analysis', *Acta Morphologica Generalis*, vol 2, no 1, pp1–8

Vengosh, A., Hirschfeld, D., Vinson, D., Dwyer, G., Raanan, H., Rimawi, O., Al-Zoubi, A., Akkawi, E., Marie, A., Haquin, G., Zaarur, S. and Ganor, J. (2009) 'High naturally occurring radioactivity in fossil groundwater from the Middle East', *Environmental Science and Technology*, vol 43, no 6, pp1769–1775

Weaver, J.L. and Asmus, D.F. (2006) 'Unitizing oil and gas fields around the world: A comparative analysis of national laws and private contracts', *Houston Journal of International Law*, vol 28, no 1, pp3–103

Worthington, P.F. (2011) 'Contemporary challenges in unitization and equity redetermination of petroleum accumulations', *Society of Petroleum Engineers Economics & Management*, vol 3, no 1, pp10–17

8 Lessons learned

Today I challenge our government and others around the world to do the exploration that needs doing. If water is in fact the new oil, let's finally do the exploration with the same vigor. We cannot begin to address sustainability issues unless we actually know how much water we have.

–Jay Famiglietti, 'Ending Our Global Water Crisis' (2013)

The following pages offer ideas that may influence the reader's approaches to managing negotiations and transforming conflicts over groundwater and aquifer use. These principles are what I have learned over the past 30 years of working with groundwater and aquifers, stumbling through my career both as a practitioner in groundwater science and engineering and as a pracademic in the world of water conflict and negotiations.

Conflicts over groundwater and aquifers are 'Jekyll and Hyde' situations. On one hand, the foundation of hydrogeology was built upon the premise of 'good' science through the precept of multiple working hypotheses, only to be corrupted by the hidden evils of junk science, stealth issue advocacy, dueling experts and the hydrohydra of myths and misconceptions within the science itself.

There is a lack of consensus among water practitioners regarding some of the fundamentals of groundwater hydrology and groundwater use sustainability. These 'skeletons in the closet' have led to the passage of public policy that is not sustainable and serves to fuel the dueling expert syndrome, as well as exacerbate the growing distrust in the opinions of groundwater professionals. Leadership and collaborative efforts to expand technical and public learning may lessen the impact of the dueling expert syndrome on the public decision-making process.

Scientists and engineers regularly encounter situations involving folk beliefs in their groundwater practice. It is unproductive to be dismissive. Discounting a believer's religion or conviction will not make the situation go away. Rather, a holistic, or transdisciplinary, approach provides a way to patiently hear the situation out and attempt to pragmatically mold what you can of those beliefs into a construct that will better serve the groundwater client's needs, yet not assail his or her belief system.

The tensions between the political and technical arenas are palpable, especially when it comes to debating how to place boundaries around the hidden resource of groundwater. Defining boundaries around groundwater resource domains is very political and polarizing because such boundaries represent different interpretations of key issues, including water quality, water quantity, nature, economics and history. When working through conflicts over boundaries, it is important to think of ways to 'blur' the boundaries by promoting the creation of a superordinate identity, emphasizing that 'we are all in this together'. The Sole Source Aquifer, hydrogeologic nature reserves and unitization provide transdisciplinary approaches to blurring the boundaries when it comes to groundwater and aquifer governance by protecting groundwater quality, quantity, stygofauna and aquifer storage. All groundwater users are shareholders in the community using an aquifer system.

Multiple uses of groundwater and aquifers extend beyond simply the commodification of water. The lack of a guiding framework for groundwater governance at this time leans toward learning how to manage the groundwater with little regard to the container – the aquifer storing the groundwater. Yet it is the aquifer conditions that affect human behavior and how groundwater is used. The chasm between the technical and the political rests on the difference between 'formal' hydrology (how scientists would like things to be) versus 'popular' hydrology (what people want). For governance of groundwater and aquifers to work, these two worldviews must merge. Integrated water resources management (IWRM) may work satisfactorily in formalized water economies, whereas in less formal water economies, institutional arrangements probably work better.

Transdisciplinary subterranean governance can enhance both groundwater and aquifer governance through accountability, thus reducing monopolies and corruption. Concurrency and unitization are emerging adaptive hybrid instruments that are both transdisciplinary in design and link the aquifer storage characteristics to groundwater use and to regular redetermination of groundwater recoverability, while looking forward to other uses of aquifers. Innovative and imaginative subsurface governance models are opening the doors to new interpretations of hydrologic and geologic information in areas previously considered water scarce. Flexible uses of transdisciplinary governance models may encourage better use of underutilized subsurface ASR (aquifer storage and recovery) and ASTR (aquifer storage, transfer and recovery) through reevaluating existing institutional policies in light of local water scarcity. Niche diplomacy in groundwater and aquifer use will increase local, regional and international stature and influence. All it will take is political leadership.

Conflict resolution and negotiation frameworks for water resources are manifold, but the use of any framework individually does not work well for groundwater and aquifers. Groundwater conflicts truly are wicked problems, and tackling wicked problems is best handled through a synthesis of the relevant conflict resolution and negotiation frameworks and a transdisciplinary imagination. Considering the tortuous journey associated with groundwater conflicts, a reasonable mental model behind the synthesis is a trifecta knot that weaves together a set of three related things that often that cause problems – the strands of water security, water diplomacy, and water conflict transformation. The 'new' siting mechanism can be used on the graduated circle of transdisciplinarity of the Hydro-Trifecta Framework in order to guide the negotiations to 'new' discoveries and paths forward.

Hydrogeologists think in four dimensions, the two dimensions associated with geography, the third dimension associated with geology, and the fourth dimension of time, typically in logarithmic cycles, given that many of the fundamental equations are based on log time. Conflict over groundwater and aquifers takes time because it takes time to build trust when discussing a hidden resource, and it takes more than an election cycle or two to observe changes in policy reflected in the aquifer system. Negotiations over developing the UN Law of Transboundary Aquifers took almost 8 years – and this was considered quick for an international instrument! In Wyoming, the aquifer protection program took 2 years to reach an agreement among nearly 10 groundwater professionals working as volunteers for a groundwater protection project, and some of the contested issues have resurfaced 15 years later. Tidwell and van den Brink (2008) report that the cooperative modeling process was held over a 9-month period for the groundwater protection project in the Netherlands, and over 18 months for the regional water quality project in New Mexico. The states of Utah and Nevada attempted to negotiate an agreement over pumping from fossil groundwater stored in the carbonate aquifers over a 4-year period, only to have the governor of Utah reject the agreement. Debates and litigation over the veracity of concurrency ordinances linking land use to groundwater availability have slowly worked their way through the courts system, only to be settled 10 years later through agreements over sharing surface water imports.

The silent revolution of small quantity wells across the western United States, Canada and many regions in the world have attempted to resolve some of the knotty disputes for the past 10 years, and despite the most well-intentioned efforts at cooperation and collaboration, these particular conflicts will probably not see resolution any time soon. The reason for the failure to reach an agreement between disputants and their respective experts and the general lack of consensus on groundwater and aquifers remains unclear, but it may have something to do with the power of the status quo. As with the idiom regarding too many cooks in a kitchen, spaghetti western water wars are ruined by the best intentions for resolving these disputes.

My community service and pro bono work on groundwater and aquifer protection taught me the importance of public participation and professional communication in the development of resource management and associated policy issues. It also underscored the importance of meshing the local history, attitudes of community members and growth and economics of the local community in the policy creation process. I observed very little in the way of integrating public participation and effective communication in the development of the revised maps, boundary delineations or suggested changes in associated policy. Local knowledge cannot be ignored. Local knowledge comes in many forms and from many sources, including homebuilders, well drillers, septic tank installers and students working on school projects. Volunteers cannot be ignored.

Academic institutions are not monolithic. Faculty and staff live in both urban and rural areas. Faculty and staff engaging in political activities are neither new nor problematic, as they have opinions as citizens. But it is also important for faculty and staff to acknowledge that their positions in the academies do not provide a bully pulpit for their opinions. The academies can provide expertise that is a solution to conflict, but also can serve as the catalyst for conflict through many avenues beyond the academies and within the academies. Water development and protection projects need public input and media coverage. A perception of bias in groundwater science or the public process poisons outcomes. A great danger exists for both science and politics when members of the scientific community participate in the 'politicization of science' through the media.

Joint fact finding is difficult to complete once trust between parties has been compromised. Middle ground solutions can be found but are time consuming and expensive. Imported expertise and consultants must be vetted for necessary licenses and knowledge of local situations in order to garner trust.

The days of groundwater problems being solved by hydrogeologists watching water move through well screens or across computer screens is quickly being replaced by the political melodramas found on the movie screen – negotiating over water use and reuse. While some may argue that documentary video could be an illusionary concept in communicating about water situations, it provides yet another avenue to enhance competencies in many different facets of professional communication. Filmmaking permits scientists to interact with the media. Individuals with little to no professional training in hydrology or documentary filmmaking successively tested the notion that facilitative documentary films can (1) promote dialogue, (2) facilitate understanding/empathy amongst parties, (3) encourage parties to consider new scenarios and (4) facilitate cooperative negotiation toward more resilient management of social-ecological systems.

Yet, the filmmaker's role is complex: The filmmaker can never be fully objective or that the camera can never be unobtrusive. The facilitative video documentarian must, out of ethical responsibility to the audience, craft a narrative. It is the filmmakers who make the documentary film rather than the subject. It is naive to promise that the documentary film will be equal and balanced for all stakeholders within the conflict.

From a pedagogical perspective, documentary videos on water situations permit practice of systems-thinking skills through situation mapping. Videos are increasingly being used as a means of community building and communication. Expect to see more discourse on water conflicts and negotiations undertaken through video as part of increasing online competency.

The 'silent pumping race' of the Disi Aquifer shared between Jordan and Saudi Arabia offers important lessons for other large groundwater development projects, such as the project anticipated by the Southern Nevada Water Authority and the carbonate aquifer shared between the states of Utah and Nevada. Both projects anticipate pumping comparable quantities of fossil groundwater. Both projects anticipate the delivery of groundwater through pipelines hundreds of kilometers in length. And, as of this writing, both projects may be operated without some form of agreement to maintain contact between the two states and two countries, so it is likely that competitive drilling and production, with consequent economic waste, physical damage to the aquifer and degraded groundwater, will ensue. It is unlikely that the pumping race between the two countries or two states will ever lead to an open conflict.

A transdisciplinary approach to new instruments of groundwater governance must not only focus on process equity and outcome equity, but also include the storage characteristics of the aquifer. Aquifer communities, concurrency and unitization provide a more holistic approach to groundwater development that acknowledges not only the pumped groundwater, but also storage of the aquifer system. The concept of 'unitization' of oil fields was developed to protect the 'corresponding rights' or 'sovereignty' of all pore space owners in the unit and to not waste valuable pore space. The result is more of a transdisciplinary, market-based approach to utilizing aquifer storage and groundwater.

The notion of unitization is not foreign to either the two countries or two states. Egypt and Yemen implement unitization in the vicinity of the Disi Aquifer. The state of Utah has experimented with the notion of unitization of groundwater in the vicinity of the Utah-Nevada border. Negotiating a unitization agreement would not necessarily require the countries or states to disclose their extraction rates if a neutral third party were involved, such as a regional mediation center or a groundwater conservation commission.

Unitization provides a transdisciplinary framework to address many other situations and benefits associated with aquifers beyond the traditional focus of groundwater allocations from aquifers. The core principles, or '4P' framework, behind the unitization of aquifers includes (1) promote groundwater exploration and development in underutilized areas, (2) preserve the storativity of aquifers, (3) private investment in technologies such as aquifer storage and recovery and (4) prevent disputes instead of conflict resolution by 'blurring the boundaries' between the two countries or states where the shared aquifer is located.

My stygian journey into the world of conflicts over groundwater and aquifers began about 15 years ago. Around this same time, noted scholar and practitioner in mediation and conflict resolution Dr. Peter Adler wrote a short piece for Mediate.com titled 'Water, Science, and the Search for Common Ground' (Adler, 2000). This article made a huge impact on me. Consider, for example, the following quote: 'Excellence in conflict resolution for water cases will derive from the way we meet the challenge of achieving powerful "substantive" solutions to tough problems. Good process and improved relationships – the traditional measures of good mediation in other arenas, are necessary but insufficient for greater use of this method in water cases. . . . In water cases, we must do better. We must be able to show outcomes that are Pareto-optimal, better than what can be achieved in litigation, better than expectations or better than some other party-established baseline'.

I interpreted Adler's commentary as underscoring the importance of meshing excellence in process *and* substance, that is, what we must do to achieve solutions to 'tough' or wicked problems is to use an integrative or transdisciplinary approach. This notion is not necessarily new to conflict resolution over ground-water and wells. Take, for example, the biblical story of Abraham who, according to Chapelle (2000), was a 'hydrogeologist' and 'well digger'. Abraham recognized that groundwater seeped from the limestone composing the hillsides. After 20 years of living in the Judean hill country, Abraham had amassed some wealth by digging wells, lining the boreholes with land stone, and artificially lifting the water for stock watering. The neighboring Canaanites quickly recognized the value of Abraham's wells and commenced taking them away. In Genesis (21:30), a dispute over the wells was negotiated, with Abraham saying, 'For these seven ewe lambs shalt thou take of my hand, that they may be a witness unto me, that I have digged this well'.

Chapelle (2000, p69) further provides some interesting commentary about the negotiations between Abraham and the Canaanites.

> When the Canaanites began to steal his wells, Abraham was presented with a nasty dilemma. He could have chosen to fight over the matter, but it probably would have cost the lives of some of his family. Certainly it would have disrupted his sheep-raising operations, and that could cause a famine. On the other hand, without his wells he couldn't raise his sheep and that could also cause a famine. Faced with these unpleasant choices, Abraham decided to see if he could work out a compromise with the Philistines. The deal Abraham struck with the Philistines cost him some sheep and oxen (Genesis 22:27), but he got his wells back. Furthermore, this treaty formally recognized Abraham's right to live in Canaan.

Abraham's situation and his approach to conflict resolution can only be described as one of the first applications of a transdisciplinary approach to conflict resolution over groundwater. Abraham applied the water security framework by acknowledging the *risk* for water use and reuse, or lack thereof, without getting into 'trouble'. Abraham also used the water diplomacy framework by addressing the *interests* for both the Canaanites and his family through a mutual gains approach to value creation. And the treaty permitting Abraham's right to live in Canaan acknowledged the water conflict transformation framework by resolving an *identity*-based conflict. In one short parable, we see that the Hydro-Trifecta Framework is not a new framework for conflict resolution for groundwater and aquifers, but rather a reframing and a synthesis of well-established, and even by some accounts ancient, water negotiation frameworks that connects well to the important questions for a transdisciplinarity resolution: What exists? What are we capable of doing? What do we want to do? And what must we do?

'As long as ground water is a source of economic prosperity, arguments over who owns it and who gets to use it are inevitable. It is not inevitable, however, that these arguments will be solved with the wisdom shown by Abraham. . . . But at least we know how' (Chapelle, 2000, p69).

References

Adler, P.S. (2000) 'Water, science, and the search for common ground', Mediate.com, www.mediate.com/articles/adler.cfm

Chapelle, F.H. (2000) *The hidden sea: Groundwater, springs, and wells*, National Ground Water Association, Westerville, OH

Famiglietti, J. (2013) 'Ending our global water crisis', speech at TEDxUCIrvine, available at www.youtube.com/watch?v=SejRgVhsT7c&feature=youtu.beTEDxUCIrvine

Tidwell, V.C. and van den Brink, C. (2008) 'Cooperative modeling: Linking science, communication, and ground water planning', *Groundwater*, vol 46, no 2, pp174–182

Appendix A
Groundwater protection role play

Acknowledgments

The following role play was developed as a supplement to the managing transboundary water resources simulation exercise developed by Len Abrams, former CEO of Water Policy Africa, currently an international consultant, water sector, in the United Kingdom for A. T. Wolf (2010) *Sharing Water, Sharing Benefits: Working Towards Effective Transboundary Water Resources Management – a Graduate/ Professional Skills-Building Workbook*, United Nations Educational, Scientific and Cultural Organization (UNESCO), Paris, France.

Disclaimer

All characters appearing in this work are fictitious. Any resemblance to real persons, living or dead, is purely coincidental.

Background

Itagatown in Itaga was awarded a grant from the Global Water Protection Agency to prepare one of the first groundwater protection programs in the world, since residents of Itagatown relied on wells and springs for over 50% of their water supply. Because of the size of the grant, Itagatown was obligated to solicit proposals from water consulting firms to complete the delineation of protection areas for the Cerulean Aquifer tapped by Itagatown's wells and springs. Itagatown felt some loyalty to EZWater, a long-established consulting firm based in Itaga, because EZWater had assisted Itagatown in preparing the grant application.

Itagatown received two proposals to complete the groundwater protection survey. Neither firm could boast more experience in completing the groundwater protection work, as the Global Water Protection Agency program was very new and no country or republic had completed this work. The proposal from EZWater cost about $500 less than the proposal from GEOWater, a newly formed consulting firm composed of specialists in groundwater engineering located in Alta in Kigala. Kigala shares the Cerulean Aquifer with Itaga and also relies on wells and springs for over 50% of its water supply. EZWater was awarded the work, based on their low price of $30,000 and their promise to complete the work in 6 months.

The groundwater scientist responsible for completing the work for EZWater was sensitive about not having finished specialty graduate work in groundwater science, but was confident that the firm's on-the-job experience for the past few years and familiarity with Itagatown's water system would provide the background needed to complete a technically and legally defensible groundwater protection report. It was important that the work be acceptable to the both Itaga and Kigala, as both countries rely heavily on groundwater. Most of the area to be delineated as a groundwater protection area was located on private property that was quickly becoming the most desirable area to live in the region.

After 1 year and after billing Itagatown nearly $50,000, the groundwater protection mapping was completed by EZWater. The beautifully colored maps showed that the most sensitive areas within the Cerulean Aquifer developed by Itagatown and Alta were located around large natural fractures called 'faults'. Most of the faults had been previously mapped by a graduate student at the Calamar Water Resources Research Institute, and the importance of faults in enhancing groundwater flow in hard rocks, such as those found outside of Itagatown and Alta, had been well documented for nearly 30 years. However, one of the large-tract landowners located in the sensitive area was concerned about the impact such a map would have on selling and developing the land. Other small-tract owners located near the mapped boundary of the sensitive area were also concerned about the development value of their land.

The first test of the groundwater protection program developed by EZWater was from a small outpatient clinic located in Calamar. The clinic would require the use of an on-site septic tank for its wastewater, as the property was located just outside the jurisdictional boundary of Itagatown and the limits of the municipal sewer lines. The property was located outside of the boundary of the groundwater protection area mapped by EZWater, but near the location of a suspected buried fault. Itagatown requested advice from the groundwater scientist working at EZWater, who indicated that, although the property was outside of the mapped protection boundary, Itagatown should be concerned that the septic tank effluent would contaminate Itagatown's groundwater and should deny the building permit.

The clinic retained the groundwater scientists at GEOWater from Alta to review EZWater's work and speak to the Itagatown Council. GEOWater determined that the clinic posed no threat to Itagatown's groundwater supply, based on recent drilling data showing that the suspected fault did not enhance groundwater flow and probably did not exist in the area of the clinic. The Itagatown Council was concerned that any successful challenges to their groundwater protection program and their consultant would undermine their efforts at groundwater protection. The landowners in Calamar and Kigala became suspicious of Itagatown's efforts to control land use outside of their jurisdictional boundaries and suspected that the groundwater protection mapping by EZWater may be technically indefensible.

With time, other technical deficiencies were uncovered in the groundwater protection mapping. Landowners within the mapped boundaries began retaining their own solicitors and technical teams to challenge Itagatown's groundwater protection program. The principal challenges to the mapped boundaries focused on the hydrologic roles of the mapped faults. Itagatown appointed a groundwater advisory committee composed of individuals trained in the legal and technical issues of groundwater. The principal advisers included the groundwater scientist who completed the groundwater protection mapping work for EZWater and the groundwater engineer for GEOWater who was based in Alta. Other advisers included the groundwater science professor at the Calamar Water Resources Research Institute, a groundwater engineering professor from the Calamar Water Resources Research Institute, the large-tract landowner and one groundwater scientist from each of the other consulting firms located in Itagatown and Alta.

The initial meeting of the groundwater advisory committee was contentious, with the EZWater consultant trying to acquire allies from the other technical advisers and with accusations of 'sour grapes' made because GEOWater was not successful in getting the Itagatown contract to map the groundwater protection areas. Sensing a story on a hotly contested local issue in the rural community, the Itagatown newspaper was in attendance. The consultant from GEOWater tried to keep the meeting on a professional level, but any attempt to start the meeting was shot down by the EZWater's constant defensive posturing. After the first 6 months of regular meetings resulted in no measurable progress, the Itagatown and Alta councils suggested alternative dispute resolution between the EZWater and GEOWater consultants.

The mediator

The mediator is a water resource professional, such as a hydrogeologist, environmental scientist or civil engineer, who has been selected due to expertise in the field and familiarity with the technical issues and jargon. Scher (1997) suggests that it is wise to seek out advisers such as a scientist or engineer with process skills and technical expertise to assist with complex environmental disputes. Case development or 'intake' will also be important, so a 'telephone call' to each disputant might provide some insight into the various issues and conflicts associated with the dispute. It is obvious to you that this dispute is oriented around many different interests and values, but that there may also be an identity-based component to it as well. On the basis of the personalities of the disputants, you suspect that to transform the conflict it might take more than one session. You are aware there are many negotiation frameworks available, so the choice is up to you.

You think there might be some value in building program development skills perhaps through a 'collaborative learning' discussion on groundwater protection using your own research, because Adler and others (2011) indicate that 'joint fact finding' and 'educating' assists the mediator in earning the trust of the disputants, as well as building trust between the disputants as they 'ask each other for help' on understanding the issues regarding groundwater or aquifer protection. Because you have little time to prepare, you decide to visit the Global Water Protection Agency site and look under Drinking Water for links to information on drinking water or groundwater protection.

During the course of completing your background research for the first meeting, you discover a typology for the boundaries for groundwater resources and user domains (Figure A.1). This work found that (1) traditional approaches to defining groundwater domains focus on predevelopment conditions, referred to herein as a bona-fide 'commons' boundary; (2) groundwater development creates a human-caused or fiat 'hydrocommons' boundary, where hydrology and hydraulics mesh; and (3) the social and cultural values of groundwater users define a fiat 'commons heritage' boundary in acknowledgement that groundwater resources are part of the 'common heritage of humankind'. Adler (2000) indicates that disputes over water are 'often large in scale, broad in impacts, and laden with values that are at odds with each other. They are emotional to both "conscience" and "beneficiary constituents". At issue in many cases are matters of culture, economics, justice, health, risk, power, uncertainty, and professional, bureaucratic, and electoral politics'. The significance of Adler's work, and the typology of boundaries that you discovered, is that it is difficult to work towards agreement without a fundamental unit of analysis. It will be important to determine what is important to the disputants with respect to the boundary issue.

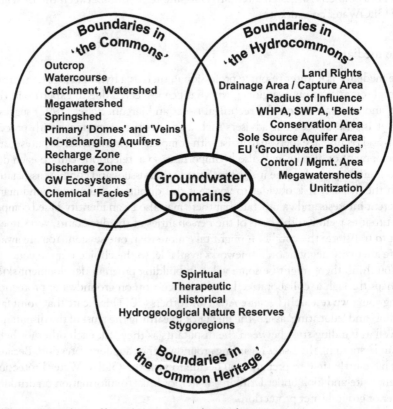

Figure A.1 Typology of boundaries in groundwater domains

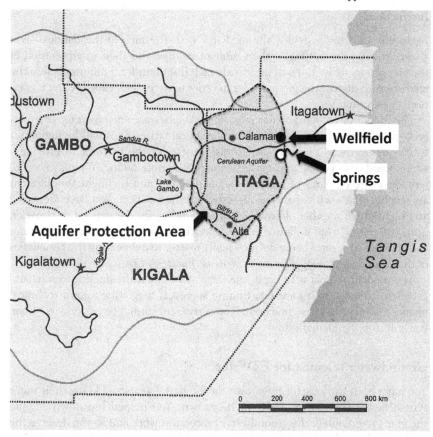

Figure A.2 Role play map for mediator; adapted from Wolf (2010)

Adler (2000) uses a quote from law professor Charles Wilkinson about how conflict transformation and the 'mediator-scientist' may serve as the 'surgeon' to begin the healing process in the Itagatown region. This quote might come in handy at the end of the first session:

> We will always have disputes over land, water, wildlife, minerals, and power. Such raspings are inevitable and ultimately healthy in a colorful, dynamic, and individualistic society. Nevertheless, the dissenting parties often leave angry, determined to undercut the temporary solution bred of combativeness. Perhaps worse, the process tears at our sense of community; it leaves us more a loose collection of fractious subgroups than a coherent society with common hopes and dreams. . . . Consensus dispute resolution involving all affected basin parties has a core value, one separate from the worth of ending a confrontation for the time being. An agreement can glue former adversaries together in a continuing process jointly conceived. Consensus builds trusting communities. Agreements heal and strengthen places.

Instructions

Each disputant has provided the mediator with the map of the Sandus River Basin showing the outline of the Cerulean Aquifer and their interpretation of the protection areas. You will meet with each disputant for 5 minutes before the mediation to discuss the ground rules in order to save time, expedite the mediation and save money.

The mediator recognizes that maps and sketches are an important part of a geologist's methods of presenting complex information, so it will be important to use an overhead projector and transparencies of the maps or to provide a white board or large pad of paper for drawing and note taking. You may want to reproduce the typology of groundwater boundaries and highlight the different approaches that each expert has suggested for the position. You have identified and highlighted another alternative for a boundary for the experts to consider, using the classic 'what if' line of inquiry, specifically as it relates to whether or not the geologic structures being debated really matter to delineating the boundaries of the groundwater quality protection areas (Figure A.2).

The mediator's goal is to break the impasse between the dueling experts and get them to agree to a joint fact finding approach suggesting a joint technical memorandum in less than one hour. If an agreement cannot be reached in 1 hour, you will stop the mediation.

Groundwater scientist for EZWater

You have been working for Itagatown for the past 2 years and know their water system better than anyone else in Itagatown. You helped Itagatown acquire the grant to complete the groundwater protection work and so you deserve the work. You have developed a strong working relationship with the water master of Itagatown.

You are almost finished with your Master of Science degree in Groundwater Science and all that is left is to finish writing your thesis. However, you are finding it hard to complete your degree with your work and the social life that has evolved since you started working at EZWater. You are aware that other groundwater scientists who have completed their advanced degrees live and work in Itagatown, but you have the same or better credentials than them with your on-the-job training, so you really don't need the degree anyway. The reason some of them are disgruntled is that you beat them on the groundwater protection proposal, and it is just a matter of 'sour grapes'.

The groundwater protection project was more challenging than you thought it would be. You spent a lot of time mapping faults that you thought were conduits for groundwater flow. You spent more time testing Itagatown's wells and observing their springs so that you fully understood how groundwater flowed to them and could better protect them. You are certain that groundwater converges on the faults, and that the faults transmit water to the wells and springs, but you just don't have the hard data to back this theory up. You used the best available information regarding the groundwater conditions in the area, including the mapping

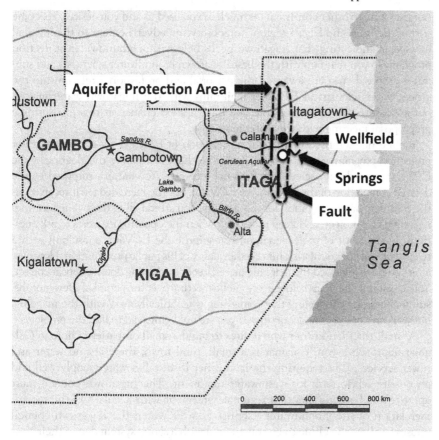

Figure A.3 Role play map for EZWater; adapted from Wolf (2010)

by the graduate student at the Calamar Water Resources Research Institute. You have to protect the reputation of your employer, EZWater, as you recently became a stockholder.

You understand that Kigala and the private landowners in the mapped area may be concerned about the development value of their land, but you just can't concern yourself with their problems. Your charge is to help Itagatown protect their drinking water supplies because they have no alternatives.

GEOWater groundwater engineer

You recently opened a branch office for GEOWater in Alta after working nearly 10 years for one of the largest groundwater engineering firms in the world. Unlike the groundwater scientist who works for EZWater, you have a graduate degree in groundwater science from Calamar Water Resources Research Institute.

You completed some work for Itagatown while living in Alta, but it was not for the Itagatown water master. You are not desperate for work, but are interested in not traveling to other continents as much as you used to and you desire to become part of the Itaga and Kigala region. You observe an advertisement in the regional newspaper indicating that Itagatown needs help with a groundwater protection program, so you team up with the best Itagatown groundwater scientists and submit a proposal. You are aware of the personal and professional relationship the groundwater scientist at EZWater maintains with the Itagatown water master, but are assured by Itagatown that you have a good chance at this job during a preproposal meeting with the water master.

You are not surprised by Itagatown's selection of EZWater for the groundwater protection program; your only regret is that you have to listen to the gloating by the EZWater groundwater scientist at regional meetings. You are not surprised by the delayed completion of the mapping by EZWater, nor the increased fee it received, as the company has a reputation for this activity. You attend all of the public meetings regarding this project, and keep to yourself when the opportunity exists to ask questions. In the spirit of professionalism, you surprise the EZWater consultant with a telephone call congratulating him on the quality of his public presentation. However, after you have reviewed the mapping more closely, something doesn't appear correct. You reexamine the maps before every public meeting over the period of a few months, still maintaining the feeling that something is technically awry with the mapping, but you can't put your finger on it, and you are not being paid to find the problems.

A small-tract landowner who desires to build a small outpatient clinic in Calamar approaches you. Calamar is a small, rural town; there are no water and sewer services, thus requiring the landowner to install a water supply well and an on-site septic tank for wastewater treatment. The Itagatown council must approve the building due to a regional agreement with Kigala. On the recommendation of the groundwater scientist from EZWater, the Itagatown council refuses to allow the clinic to be constructed with a septic system due to their concerns over impacts on the Cerulean Aquifer, which Itagatown uses for its drinking water supply. This position surprises you, as the land in question is outside of the mapped boundary for the groundwater protection program that Itagatown is implementing. You are requested by the landowner in Calamar to present the facts and findings of a technical review of the groundwater protection program to the Itagatown council. You point out that the land is outside the mapped boundary, and that based on the results of drilling the water supply well for the clinic, the local hydrology is favorable for protecting the groundwater supplies of Itagatown. The EZWater consultant indicates that while the subject property is outside the mapped boundary of the groundwater protection area, a buried fault is located close to the property that places Itagatown's drinking water supplies at risk. You provide additional information detailing why you believe the fault does not exist, but are unsuccessful in making your case.

Other landowners in both Itaga and Kigala and their technical consultants also encounter the same treatment by Itagatown and the EZWater consultant. You continue to attend the public meetings and listen to the same complaints, as well as the answers to those complaints. In your discussions with other consultants, you discover that there is interest in redoing the groundwater protection program using volunteers from other firms, the primary landowners and instructors at the Calamar Water Resources Research Institute.

The Itagatown and Alta councils request your technical assistance with coordinating the volunteer efforts to redo the groundwater protection mapping project. You are concerned that any changes made to the mapping will be perceived by the EZWater consultant as a personal attack on the consultant and the firm, or just 'sour grapes' because GEOWater was not selected to complete the work. You are also concerned about the perception of uncertainty and indecision in the groundwater sciences by the public. You are also concerned that, while the groundwater protection program is important to every community in the region, it will not be embraced by other water systems because it could not be developed with any consensus in the very region that is the technical center of excellence for groundwater science in the world. One thing is certain, however: the mapping completed to date is technically impracticable because it requires groundwater to move uphill towards the faults, which is contrary to the previous graduate student's work and contrary to the fundamentals of groundwater science in that groundwater flow obeys the laws of gravity – by flowing from areas of high hydraulic head or elevation to areas of low hydraulic head or elevation. You have kept this observation quiet for some time, hoping to get the EZWater consultant to make concessions on his work without embarrassing him in front of his peers and the Itagatown council.

You have developed an alternative method to mapping the groundwater protection zones that has been used on another continent; this method has been informally discussed with the landowner in Calamar, as well as with the large tract landowners in both Kigala and Itaga and other regional consultants, who all agree that it will probably work for the Cerulean Aquifer. The technical approach is simple and does not rely on the ambiguities of deciding which fault enhances groundwater flow and which fault has no effect on groundwater flow. The technical approach relies on assessing whether Itagatown and Alta develop groundwater from the entire Cerulean Aquifer or just discrete parts of the Cerulean Aquifer. If Itagatown only relies on one part of the Cerulean Aquifer, then the mapping approach should focus on that part of the Cerulean Aquifer. If there is concern that protecting one part of the Cerulean Aquifer does not look out for the future supplies of Itagatown, then the currently developed portion of the Cerulean Aquifer should receive priority protection as designated on a map, while acknowledging that the other portions of the Cerulean Aquifer are important for the future of Itagatown's and Alta's water supplies, but receiving a lower priority on a map. The technique recognizes the uncertainty of the hydrologic role of faults and recommends that decisions regarding the risk of developing land near a mapped or suspected buried fault be based on site-specific drilling investigations paid for by the party interested in developing the land. The technique also recognizes the importance of the rivers that cross the Cerulean Aquifer, where snowmelt and stormwater from the mountains located in the western part of Kigala, Gambo and the Sandus Republic converge and are observed to 'disappear' and recharge the aquifer.

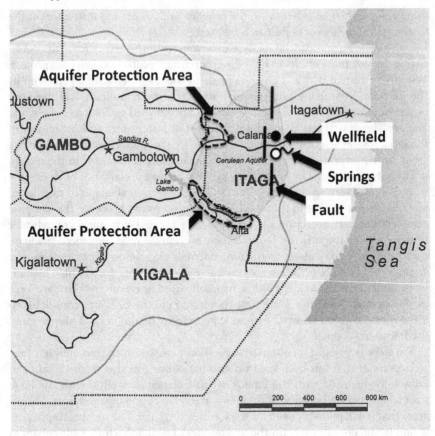

Figure A.4 Role play map for GEOWater; adapted from Wolf (2010)

Integration of this technical approach recognizes the potential importance of faults in transmitting groundwater to some wells and springs, as reported by the EZWater consultant, but relies on a more holistic approach to groundwater protection. Your interest is designed to salvage the project in order to gain the trust of the public in the value of groundwater sciences in their everyday lives, show other water systems in the state that the mapping of groundwater protection areas can be technically defensible using volunteers and stop providing public funds (your tax dollars since you are also a resident) to your competitor.

References

Adler, P.S. (2000) 'Water, science, and the search for common ground', Mediate.com, www.mediate.com/articles/adler.cfm

Adler, P.S., Bryan, T., Mulica, M. and Shapiro, J. (2011) 'Joint fact finding a strategy for bringing science, policy and the public together when matters get contentious', Mediate.com, www.mediate.com/articles/AdlerJoint.cfm

Scher, E. (1997) 'Using technical experts in complex environmental disputes', Mediate.com, www.mediate.com/articles/expertsC.cfm

Wolf, A. T. (2010) *Sharing water, sharing benefits: Working towards effective transboundary water resources management – a graduate/professional skills-building workbook*, United Nations Educational, Scientific and Cultural Organization (UNESCO), Paris, France

Appendix B
Listing of water documentary videos

A Journey in the History of Water (2001) is a documentary based on the television series *History of Water*. It is a four-part series: 'The Struggle', 'The Energy', 'The Myths' and 'The Conflicts'. Each segment is 45 minutes long. Norwegian Broadcasting Corporation. http://watervideo.com/video.htm

A Regional Love Story – Friends of the Earth, Middle East (2011) is a short video that reflects the insights and learning outcomes of the 'Water Trustees': Jordanians, Palestinians and Israelis who live along the Jordan River. Ideas and messages in song were created by Regional Youth Friends of Earth Middle East. Joseph, who is the coordinator Jordanian poet, collected the content shared in groups and wrote lyrics that seem to emanate from the mouth of the River Jordan itself. The Hebrew lyrics and melody were written in the Yishai Oz joint workshop with the youth of the three nations. The producers comment that they hope the work together in the 'Good Water Neighbors' project will contribute to the rehabilitation of the Jordan River in the south and the improvement of relations between the peoples residing along its waters. youtube/hi06Q9VNS7c

A River Loved: A Film About the Columbia River & the People Invested in Its Future (2011) is a documentary film that tells the story of the Columbia River and the diverse people and interests in the basin. The Columbia River has been successfully managed by the United States and Canada for hydropower and flood control since the 1960s. The Columbia River Treaty is an inspirational example of international cooperation; however, needs and values for the basin have changed since the 1960s. 37 minutes. Funded in part through the US Geological Survey 104(b) program. Available free on YouTube (youtu.be/ZTIj8zIugdA).

Battle for the Klamath (2005) is a feature-length documentary about the fight over water rights and salmon between environmentalists and Indian tribes on one side and small farmers and the Bush Administration on the other. 56 minutes. Veriscope Productions. www.veriscope.com/store/

Blue Gold: World Water Wars (2008) is a based on the book, *Blue Gold: The Right to Stop the Corporate Theft of the World's Water*, by Maude Barlow and Tony Clarke. PurpleTurtle Films, www.bluegold-worldwaterwars.com/. *Bull Run* (2009) is a film about the Bull Run River, which has its source in the foothills of Mt. Hood and has supplied Portland's drinking water since 1895. To protect the water from contamination, the upper watershed has been closed to the public for more than a century. 29 minutes. Oregon Public Broadcasting, Portland, OR 97219

Cadillac Desert: Water and the Transformation of Nature (1997) is a four-part documentary series about water, money, politics and the transformation of nature. This series more or less started the water documentary genre based on Marc Reisner's book, *Cadillac Desert* (1986) and Sandra Postel's book, *Last Oasis* (1992). A Public Broadcasting Service classic. www.ldeo.columbia.edu/_martins/hydro/case_studies/cadillac_desert.htm

Celilo Falls and the Remaking of the Columbia River (2006) examines a turning point in the history of the Pacific Northwest. For millennia, Celilo Falls was the great Indian fishery on the mid-Columbia, and it drew Indians there from throughout the West to trade for salmon. But in 1957, the federal government began operation of a giant hydroelectric dam at the Dalles that drowned Celilo Falls and ended the fishery there. 29 minutes. Oregon State University, Corvallis, OR 97331

Connecting Delta Cities (2009) is a documentary that explores the impact of global warming on coastal cities such as New York, Jakarta, Rotterdam and Alexandra. The film shows that all coastal cities face similar problems that can be tackled more efficiently through an exchange of knowledge and intensive cooperation. Two versions available – 17 minutes and 42 minutes. Rodes Vis Producties. www.connectingdeltacities.com

Downwind Downstream (1988) documents the serious threat to water quality, subalpine ecosystems and public health in the Colorado Rockies from mining operations, acid rain and urbanization. 60 minutes. Environmental Research Group. Bullfrog Films, Oley, PA 19547

Drowned Out (2002) is about an Indian family who decide to stay at home and drown rather than make way for the Narmada dam. This documentary follows the Jalsindhi villagers through hunger strikes, rallies, police brutality and a 6-year Supreme Court case. Two versions – 75 minutes and 45 minutes. Spanner Films. www.spannerfilms.net. Bullfrog Films, Oley, PA 19547

FLOW: For Love of Water (2008) investigates what experts label the most important political and environmental issue of the 21st century – the World Water Crisis. The director builds a case against the growing privatization of the world's dwindling fresh water supply with an unflinching focus on politics, pollution, human rights and the emergence of a domineering world water cartel. 84 minutes. www.flowthefilm.com/. Oscilloscope Laboratories (www.oscilloscope.net) Nestle Waters North America response to 'FLOW: For Love of Water' 10 minutes. www.nestlewatersvideos.com/responses/flow_response/

FrackNation (2013) is a feature documentary that counters *Gaslands* about fracking for natural gas. 76 minutes. Hard Boiled Films. fracknation.com/

Gasland (2010) and *Gasland, Part II* (2013) are documentaries looking at the dangers of hydraulic fracturing, or fracking, the controversial method of extracting natural gas and oil, now occurring on a global level (in 32 countries worldwide). Part II argues that the gas industry's portrayal of natural gas as a clean and safe alternative to oil is a myth and that fracked wells inevitably leak over time, contaminating water and air, hurting families and endangering the earth's climate with the potent greenhouse gas, methane. 107 minutes. HBO Documentary Films www.gaslandthemovie.com/home

Last Call at the Oasis (2011) sheds light on the vital role water plays in our lives, exposes the defects in the current system, shows communities already struggling with its ill effects and introduces us to individuals who are championing revolutionary solutions, such as activist Erin Brockovich and distinguished experts Peter Gleick, Alex Prud'homme, Jay Famiglietti and Robert Glennon. 105 minutes. Docuramanfilms. docurama.com

Liquid Assets: The Story of Our Water Infrastructure (2008) tells the story of essential infrastructure systems: water, wastewater and stormwater. These systems – some in the ground for more than 100 years – provide a critical public health function and are essential for economic development and growth. Locations featured in the documentary include Atlanta, Boston, Herminie (Pennsylvania), Las Vegas, Los Angeles, Milwaukee, New York City, Philadelphia, Pittsburgh and Washington, DC. 90 minutes. Penn State Media, University Park, PA, liquidassets.psu.edu

Nuestras Acequias (2004) follows the acequias as they weave down the hills and across the fields of Northern New Mexico and provides the stories of villagers linked by these ditches. 20 minutes. Rivers & Birds, Arroyo Seco, NM 87514

Red Gold (2009) is a documentary about the Bristol Bay region of southwest Alaska, which is home to the Kvichak and Nushagak rivers, the two most prolific sockeye salmon runs left in the world. Two mining companies, Northern Dynasty Minerals and Anglo American, have partnered to propose an open-pit and underground mine at the headwaters of the two rivers. 55 minutes. Felt Soul Media, feltsoulmedia.com

Rising Waters: Global Warming and the Fate of the Pacific Islands (2000) uses personal stories of Pacific Islanders in Kiribati, the Samoas and the atolls of Micronesia, including the Marshall Islands and Hawaii, as well as researchers in the continental United States to put a human face on the international climate debate. Already, unusually high tides have swept the low-lying atolls of Micronesia, destroying crops and polluting fresh water supplies. 57 minutes. Torrice Productions. Bullfrog Films, Oley, PA 19547

River of Renewal (2008) tells the story of conflict over the resources of California and Oregon's Klamath Basin. Over the years, different dominant groups have extracted its minerals, trees, and water with disastrous consequences, including the collapse of industries and of wild salmon populations. This film documents protest and acts of civil disobedience as Indian tribes, farmers and commercial fishermen defend their ways of life. It witnesses a remarkable turnaround as politically polarized stakeholders and government agencies overcome bitter divisions in reaching a consensus about the future. 55 minutes. Pikiawish Partners Production, www.riverofrenewal.org/

Riverwebs (2007) follows several international researchers through a true story of personal tragedy, growth and recovery, providing a glimpse into the hidden world of rivers and the scientists who explore them. The film chronicles the inspiring life and work of the pioneering Japanese ecologist, Dr. Shigeru Nakano. 57 minutes. Freshwaters Illustrated, www.freshwatersillustrated.org

Running Dry (2005) is a compelling documentary about water quality and quantity around the world. The film gets behind the headlines and superficial treatments to grapple with the complexities of the global water crisis. 52 minutes. The Chronicles Group, www.runningdry.org

Source to Sea: The Columbia River Swim (2006) takes a broad, sobering look at the history and destiny of what the indigenous people call 'Che Wana', the Columbia River. Full of rare archival footage of the sacred Celilo Falls and Kettle Falls, now inundated by dams, the film also deals with the Hanford Nuclear Reservation, pollution from mining, the river in Canada, petroglyphs, salmon issues and the struggles Christopher Swain encountered. 90 minutes. Gryfalcon Films, LLC Portland, OR.

Tapped (2010) is an inspiring documentary that trails the path of the bottled water industry from the plastic production to the ocean in which so many of these bottles end up. 76 minutes. Gravitas Ventures LLC, www.tappedthemovie.com/

The American Southwest: Are We Running Dry? (2008) provides a definitive look at how the water crisis affects the American Southwest states. The film represents a profound opportunity to educate the public about conservations, water reuse, desalination, population growth and future water policies. 71 minutes. The Chronicles Group, www.runningdry.org

The Future of Water (2011) tells the story of how the struggle to control and use water will have a great impact on political power relations worldwide and influence war and peace and the destinies of countries and entire continents. The film has three episodes: 'The Waterlords', 'The New Uncertainty' and 'The Water Age'. 150 minutes. http://watervideo.com/future/

The Oregon Story: Water (2003) examines what is perhaps our most precious resource. Most of us think of Oregon as having abundant water supplies, but that is not the case. This documentary explores both the history of water allocation around the state and how society's demands upon it have changed over time. You'll see examples of water-based problems and conflicts and meet some of the people who are leading the way in efficient usage and water conservation. 60 minutes. Oregon Public Broadcasting, www.opb.org/programs/oregonstory/water/program.html

The Unforeseen (2007) is an urgent, beautifully crafted documentary in which the American dream of owning a house with a white picket fence goes head to head with environmental sustainability. When an ambitious real estate developer sets out to transform thousands of acres of pristine hill country in Austin, Texas into a suburban development – threatening a nearby natural spring – the community fights back. 93 minutes. Ojo Partners and the Cinema Guild, www.cinemaguild.com

Thirst (2004) focuses on the control over public water supply – a significant policy issue around the world. This documentary takes a close look at the global business trend of privatizing water supplies, while addressing the question, Is water a human right or a commodity to be bought and sold in a global marketplace? 65 minutes. Bullfrog Films, Oley, PA 19547, www.thirstthemovie.org/

Unconquering the Last Frontier (2002) chronicles the historic saga of the damming and undamming of Washington's Elwha River. The film tells the story of the Elwha Klallam tribe's struggle to survive in the shadow of hydroelectric development. The federal government has given approval to Elwha River restoration, in the largest scale dam decommissioning project in the world. 57 minutes. RLA, LLC. Bullfrog Films, Oley, PA 19547

Upstream Battle (2008) depicts Native Americans' fight for the survival of their salmon and their culture – against an energy corporation. Their struggle may trigger the largest dam removal project in history. Three versions available – 26, 59 and 97 minutes. Preview Production GbR, www.upstreambattle.com/

Upstream/Downstream (2003) is a documentary funded by the Asian Development Bank and part of ADB's Water Voices documentary series. It tells the story of how various stakeholders of the Thailand's Ping River resolve conflict over water use. About 22 minutes. Part of the Water Voices Documentary Series of compelling stories about people tackling water problems in Asia and the Pacific, by Halsey Street Ltd., New Zealand. Available free on YouTube, www.youtube.com/watch?v=t8n7pPwM9kM&feature=share&list=PLFFB081288A7EC0C3

Water Before Anything: Crisis and Transformation (2009) is an engaging film made by an Oregon State University master's student looks at the possibility of water as a force for bringing people together – rather than pushing them apart. We are invited into a small community in Oregon, where over the course of 5 years, residents worked together to find solutions to their water crisis. The filmmaker interviews a range of stakeholders on the hardships and hopes for their community and integrates these elements with the science of groundwater and the complex steps of a regional task force working through conflict resolution to form policy. About 40 minutes. Funded through the U.S. Geological Survey 104(b) program. Available free on TheWaterChannel, www.thewaterchannel.tv/index.php?option=com_hwdvideoshare&task=viewvideo&Itemid=4&video_id=1072

Water First: Living Drop by Drop (2006) is a documentary made in Malawi that highlights the central importance of clean water in relieving poverty and empowering people. Through the inspiring story of Charles Banda – a humble, local Malawian fireman turned waterman who has provided clean water to hundreds of thousands of fellow Malawians – this documentary is a personal story with global significance. 25 minutes. Hart Productions, New York, NY www.waterfirstfilm.org/data/

Water's Journey: The Hidden Rivers of Florida (2003) is a documentary film that tracks the path of water through the Floridian aquifer, where a team reveals the journey of water above and within the earth. Viewers are transported through a world that reveals how their lives are intertwined with the water they drink. 57 minutes. Karst Productions, Inc. Available free on www.floridasprings.org/expedition/videos/

Waterbuster (2006) covers the story of a Hidatsa/Mandan filmmaker as he revisits the Upper Missouri River basin in North Dakota where his ancestors once lived. There, he investigates the impact of the massive Garrison Dam project, constructed in the 1950s by the U.S. Army Corps of Engineers, which laid waste to a self-sufficient American Indian community, submerging fertile land and displacing the filmmaker's family and the people of the Fort Berthold Indian Reservation. 78 minutes. Brave Boat Productions, Inc. Quechee, VT 05059

Waterlife (2009) is a documentary on the extraordinary beauty and complex toxicity of the Great Lakes following the epic cascade of its waters from Lake Superior to the Atlantic Ocean. 109 minutes. Primitive Entertainment and the National Film Board of Canada, waterlife.nfb.ca. www.ourwaterlife.com/index. html

Wind River (2000) is a modern day story of cowboys and Indians. White rangers on the Wind River Indian Reservation are fighting to protect the right to water for irrigated agriculture. The Shoshone and Northern Arapaho tribes are fighting to save the de-watered Wind River as a part of their own heritage. 34 minutes. High Plains Films. Bullfrog Films, Oley, PA 19547

References

Postel, S. (1992) *The last oasis: Facing water scarcity*, W. W. Norton & Company, New York, NY

Reisner, M. (1993) *Cadillac desert: The American west and its disappearing water*, revised and updated, Penguin Books, New York, NY

Index

Printed in the United States
by Baker & Taylor Publisher Services

the United States
Taylor Publisher Services